BASIC TENETS OF
RELATIVISTIC
ASTROPHYSICS
AND
COSMOLOGY

BASIC TENETS OF
RELATIVISTIC
ASTROPHYSICS
AND
COSMOLOGY

A R Prasanna

Physical research Laboratory, Ahmedabad, India

World Scientific

NEW JERSEY · LONDON · SINGAPORE · BEIJING · SHANGHAI · HONG KONG · TAIPEI · CHENNAI · TOKYO

Published by

World Scientific Publishing Co. Pte. Ltd.
5 Toh Tuck Link, Singapore 596224
USA office: 27 Warren Street, Suite 401-402, Hackensack, NJ 07601
UK office: 57 Shelton Street, Covent Garden, London WC2H 9HE

Library of Congress Control Number: 2024950374

British Library Cataloguing-in-Publication Data
A catalogue record for this book is available from the British Library.

ISBN 978-981-98-0845-8 (hardcover)
ISBN 978-981-98-0846-5 (ebook for institutions)
ISBN 978-981-98-0847-2 (ebook for individuals)

For any available supplementary material, please visit
https://www.worldscientific.com/worldscibooks/10.1142/14191#t=suppl

Typeset by Stallion Press
Email: enquiries@stallionpress.com

Dedicated to the memory of
Professor Jayant V Narlikar
a source of inspiration

Preface

Few years back, I received a letter from a university Physics department requesting me to give a talk to graduate students particularly on aspects of physics in astrophysics, to get them motivated to study the subject of astrophysics. I took the opportunity to prepare a course of topics highlighting the aspects of astronomy and physics along with basic ideas of some modern aspects in then currently dealing with research in astrophysics. Having prepared the material I realized that I had some material to bring out a monograph on the basic tenets of relativistic astrophysics and cosmology that resulted in this little book.

It starts with recollecting some of the most beautiful pictures of the cosmos that have enamoured humanity and sketching a brief history of the birth and death of stars. On the theoretical side the significance of the main pillars of the twentieth-century physics-Quantum theory and theory of Relativity need to be reemphasized citing their requirement in understanding the cosmos. On the observational side, the discovery of radio frequency emitting objects in the forties added immensely to our viewing of the universe that resulted in the study of radio astronomy which is now a more dominant partner. Since the universe is made up of space, time and matter, it is very pertinent to introduce the theory of general relativity that became an important basis for the understanding of the energetics of cosmic objects and provide a basic structure for cosmology.

Going from basic features to a little advanced aspects, the discussion takes one through the aspects of gravitational collapse which deals with the story of stars ending their lives as compact objects and blackholes.

As mass and charge are the most dominant properties of matter, the forces associated with gravitation and electromagnetism need to be discussed in consonance depending upon their dominance. In the cosmos where matter dominates gravitation provides the basic geometry as explained by general relativity and electromagnetism needs to be considered on this background geometry. The monograph looks on this aspect by dealing with particle trajectories and discussion of electrodynamics, hydrodynamics, and plasma physics in a general relativistic formalism. The energetics of blackholes takes one to the discussion of the process of accretion and the associated physics. After citing several different studies concerning accretion both spherical and disk, the discussion moves on to the two important studies — gravitational lensing and gravitational radiation (waves).

Finally, the story of cosmology deals with the very early studies of Einstein and Friedmann and moves on to describe various cosmological models that form the basis for understanding the structure of the universe. After a brief survey of various cosmological models, the discussion considers briefly the early universe scenario of hot big bang theory, its implications leading to a mention of the dark matter or the missing mass and finally the notion of dark energy.

The entire presentation is descriptive without being rigorous and more on the lines of creating sufficient interest for the reader to look at the abundant literature on these topics that has accumulated over the past fifty years.

I wish to acknowledge the help I have received from the Physical Research Laboratory, Ahmedabad, its Director and most importantly the library staff who have been immensely cooperative in finding the references whenever I approached. Some of the chapters do have material consisting of my own work with colleagues and students over the past several years for which I want to express my sincere thanks to all of them. I would like to place on record the very valuable help received from the corresponding editor Ms Carmen Teo Bin Jie and her team for getting the ms ready in final form for production. My wife Shanti has been very understanding during the saga of writing and I express my appreciation to her undivided support.

Contents

Chapter 1

Introduction

All matter and radiation that pervades space and time is endowed with motion and thus is spread over the vast expanses of the universe. The study of space, time and its contents with their motion is referred to as cosmology. The main force or interaction that keeps this motion is gravitation which as described by Newton is universal and sustains mass and energy of the cosmic distribution. It is quite well known that though initially, before the arrival of Copernicus, it was thought that Earth was the centre of the Universe but with Copernicus, the heliocentric theory got established that explained the motion of planets around the Sun by Kepler, and with time came the realisation of the motion of Sun in the galaxy, and of galaxy in the local cluster of galaxies and of the cluster in the super cluster so on and on. This vast collection of matter and associated structures how did they form and how are they holding up has been a subject that has occupied the human mind for a long time. However, after the Copernican revolution brought in the heliocentric system of reference to understand the Universe, attempts have been initiated to describe the structure of the Universe and its contents from macro to micro a study that is in general referred to as "Astrophysics".

One of the first noticed views of the Universe, apart from the happenings of day and night and seasons is, being (looking) the same in all directions over large distances and for all observers over eons of time. Geometrically these features are referred to as universe being homogeneous and isotropic.

This in fact became the first assumption for constructing any model of the Universe. These two features of isotropy and homogeneity are actually based on the fact that every observer on Earth is in constant motion in space as there exists constant motion of earth around its axis, as well as around the sun, of sun in the Galaxy, of Galaxy in its cluster and so on as mentioned earlier and thus an observer's position in the Universe keeps changing continuously but yet the view of the Universe remains the same. However, as this systematic motion on a large scale of structures that are major constituents was not observable before the invention of the telescope, most astronomers of the nineteenth and early twentieth centuries assumed the Universe to be static. As new instrumentation and techniques were developed in the mid-twentieth century, including satellite technology it became possible to view the universe in the entire electromagnetic regime from radio to gamma rays, which offered deeper and better understanding of the universe. In this understanding, the main role was due to the application of developments both in macro and micro physics along with associated development in mathematics

1.1 Recalling Some Known Facts

Improving upon the invention of a telescope by the Dutch optician Lipperhey, Galileo (using lens) and Newton (using mirrors) improved the instrument to look at the sky and the visible cosmic objects. These efforts revealed the existence of Jupiter's moons, Saturn's rings the outer planets Neptune, Uranus and the Milky way galaxy.

With improved technologies, one could see a vast majority of objects like galaxies, nebulae, stellar and galactic clusters, and several similar cosmic objects. This amazing view of the Universe kept the human mind occupied thinking about what, how and why of the world around us. With the advent of space technology and the launching of a 2.4 m optical telescope (Hubble) in the nineties, the viewing increased by a thousand fold and one now has an incredible number of photographs and spectrographs of distant cosmic objects up to the distance of about 13.4 billion lys. The most novel feature of the enlarged view is the ability to distinguish several individual stars and stellar systems and get their spectrographs which

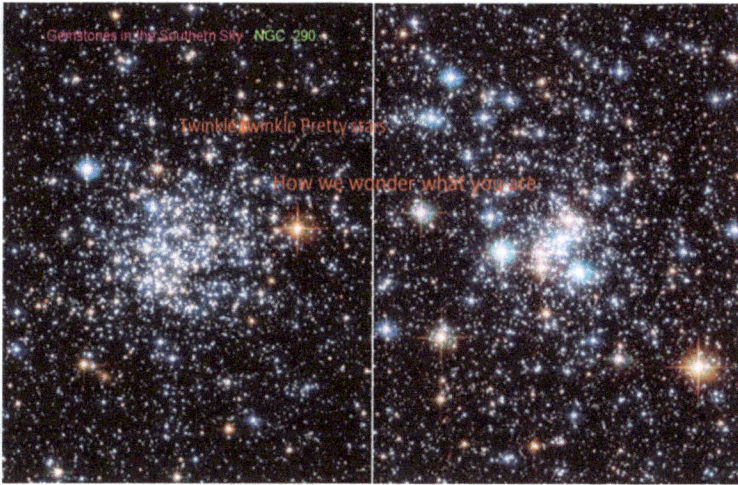

Fig. 1.1. *Image Credit: NASA/ESA/STScI.*

Fig. 1.2. A montage of Jupiter and its four largest moons (distance and sizes not to scale). *Image Credit*: Wikipedia, the free encyclopedia creative commons.

reveal their motion and also the nature of their material content. Some examples of such objects are depicted in Figures 1.5–1.7.

Recently with the arrival of the J. Webb multi-mirror telescope (launched in Dec 2021) and located at the second Lagrange point between

Fig. 1.3.   The Sun and nine planets. *Image Credit*: NASA, Kelvin J. Hamilton.

Fig. 1.4.   The Milky Way galaxy. *Image Credit*: ESO/S. Brunier.

Earth and Sun, the viewing capacity has increased with the ability for higher resolution than the Hubble telescope and one expects to learn much more about the Universe. The location of this telescope about 1.5 million kilometres away from the Earth is very appropriate for extending the viewing from the optical window to far infrared and infrared windows as the temperature at the telescope site is extremely low (39 K).

Fig. 1.5. Andromeda Galaxy. *Image Credit*: NASA/JPL-Caltech.

Fig. 1.6. The Horse head Nebula. *Image Credit*: CFHT, Coelum, MegaCam, J.-C. Cuillandre (CFHT) & G. A. Anselmi.

Fig. 1.7. Pin Wheel galaxy. *Image Credits*: Hubble Image: NASA, ESA, K. Kuntz (JHU), F. Bresolin (University of Hawaii), J. Trauger (Jet Propulsion Lab), J. Mould (NOAO), Y.-H. Chu (University of Illinois, Urbana) and STScI; CFHT Image: Canada-France-Hawaii Telescope/J.-C. Cuillandre/Coelum; NOAO Image: G. Jacoby, B. Bohannan, M. Hanna/NOAO/AURA/NSF.

Before these developments, K. Jansky and G. Reber had added the radio window for the Universe in the 1930s which aided and continues to add an ever-increasing variety of cosmic objects to the celestial zoo with radio frequency emission. With the discovery of extremely energetic sources (of the order 10^{52} erg/sec, being equivalent to 10^{46} Joule/sec) like the Active Galactic Nuclei and Quasars in the 50s and early 60s, and the incredible discovery of pulsating radio sources the Pulsars (Bell 1967). Radio astronomy gained a prime position in astrophysical research. Further, as will be discussed, the counting of radio sources as carried out by M. Ryle (1956) had important input for deciding as to which of the cosmological models gives a better description of the Universe.

Apart from the Radio, Optical, FIR, and IR windows one other low-frequency source for astronomical studies comes from the Microwaves and a very important input was from the accidental discovery of the cosmic microwave background radiation (CMBR) by Penzias and Wilson in

Fig. 1.8.   Quasar 3C 273 taken by Hubble Space Telescope. *Image Credit*: NASA/ESA.

Fig. 1.9.   Computer drawing of galactic core credit National Radio Astronomy Observatory, California Institute of Technology; Walter Jaffe/Leiden Observatory, Holland Ford/JHU/STScI, and NASA.

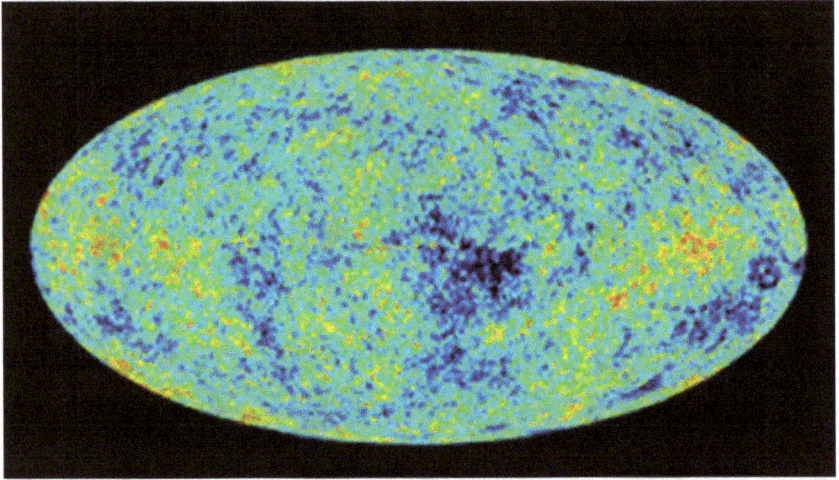

Fig. 1.10. Cosmic Microwave Background WMAP. *Image Credit*: Shu, Frank H., "cosmic microwave background". *Encyclopedia Britannica*, 20 Mar. 2023, https://www. britannica.com/science/cosmic-microwave-background. Accessed 29 April 2023.

1965 which was originally thought to be homogeneous and isotropic pervading the entire universe but later found to have anisotropies as shown in the picture above.

The most important source of energetics for these cosmic objects was earlier thought to be only nuclear burning which was later found to be inadequate with the discovery of active galactic nuclei and quasars. Hoyle and Fowler (1963) proposed the possibility of the gravitational collapse of heavy mass stars which could release the gravitational potential energy amounting to high intensity emissions. An important factor realised in this context was that the Newtonian description of gravity was not sufficient to explain the energy and one requires the general relativity of Einstein to describe the scenario.

While observations are the most important aspect of understanding the Universe, one needs to clearly understand the science of the processes that lead to make these objects visible or recognizable which is achieved through studying the physical processes that lead to the production and emission of radiation from cosmic objects.

The most important developments in modern physics (physics of the twentieth century) that are essential for this analysis are the Quantum

Physics of Max Planck (1900) and the Special theory of relativity of Albert Einstein (1905). Both these theories have played prominent roles in the understanding of energy production and emission mechanisms from astrophysical objects. It is well known that Gravity is the most important interaction in the universe that governs every particle and while the Newtonian description of gravity helps one to understand its role on earth, and in the solar system, the description of the vast universe and its contents with massive bodies requires Einstein's theory of General relativity that describes the role of gravity as curvature of space-time and its significance in analyzing the trajectories of test particles and their dynamics.

While the later half of the last century pushed forward the developments and a deeper understanding of the observations in the electromagnetic window, a very significant discovery was made at the beginning of the present century — the detection of 'gravitational waves' (2015) which has increased the impact of astrophysics and cosmology as instrumental in developing the role of physics for understanding the Universe. Einstein, after describing gravity as the curvature of the associated space-time geometry had predicted in 1917 through solving the linearized form of the field equations in the weak field limit, the possible existence of gravitational waves which are similar to the electromagnetic waves associated with the electromagnetic field. As both these fields are of far-field nature (both interactions follow the inverse square law $1/r^2$), the waves associated with these fields are very significant in carrying out the information about their sources as well as of the intervening medium through which they propagate. Detection of gravitational waves in 2015 further enhanced the importance of general relativity to describe the physics of the universe and its contents. On the other hand Hubble's discovery of the velocity–distance relation of galaxies and clusters had mooted the idea of the expanding universe which again needed the understanding of space-time physics as required by general relativity for the study of Cosmology.

Theoretical understanding in these areas required a large input from the studies of the theory of elementary particles which has been an integral part of the physics of the cosmos almost since the early 1970s, when lack of experimental facilities for increasing the energy ranges of accelerators that could be built, resulted in a large influx of particle physicists into investigations in high energy astrophysics and early cosmology where the role of elementary particles take the center stage. Added to these were the studies in neutrino physics which obviously was associated with the detection of neutrinos from cosmic sources (particularly the solar neutrinos) as a result of various physical processes expected to occur in varied astrophysical interactions. As an overall view, one can see that the astrophysics of the last hundred years has initiated and developed most of the studies both in experimental and theoretical aspects of physics research and continues to do so as the areas of investigations include analysis and description of all the four fundamental interactions of nature lead by Gravity and electromagnetism followed by the two short-range forces, the weak interactions and the strong interactions. Initially, the sources for observational astronomy were in the different frequencies of the electromagnetic spectrum, from different sources. However, the detection of gravitational waves coming from the coalescence of two neutron stars in 2017, along with the associated detection of signals in few bands of the electromagnetic spectrum-γ-rays, X-rays, ultraviolet, optical and infrared, from the same region of space gave birth to multi-frequency astronomy and a host of important physics related questions for explaining their emissions and their propagation. The scenario today regarding research in astrophysics and cosmology is the most fertile ground for physicists as it brings in the features of almost all aspects of Hydrodynamics, Electrodynamics, Thermodynamics, Plasma physics, high energy physics and the associated mathematics, making it a very rich discipline to take up for investigation, discovery and enjoyment of research.

1.2 Birth of a Star

The Universe is filled with the most abundant element Hydrogen in the form of molecular clouds. Initially, as there is no source of heat, the clouds contract and gravity the universal attractive force brings the atoms of the hydrogen together and they start fusing with one another like four atoms of hydrogen combining to form an atom of Helium the next element in the periodic table. This process of Helium formation from the fusing of deuteron atoms in stars was first discussed by Hans Bethe, a well-known nuclear physicist (1939). The energy production of sun-type stars is due entirely to the combination of four protons and two electrons into an α-particle which is called the 'p-p reaction'. Two protons interact to form Deuteron which then captures two more protons rapidly to form He^4. Similar ideas were also expressed by Gamow and Weizsacker around the same time. In fact, Bethe had suggested one more possible reaction with the same probability, a chain reaction that uses carbon and nitrogen as catalysts commonly known as CNO cycle:

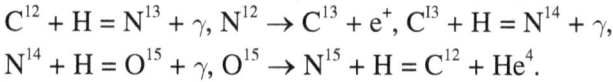

$$C^{12} + H = N^{13} + \gamma, N^{12} \rightarrow C^{13} + e^+, C^{13} + H = N^{14} + \gamma,$$
$$N^{14} + H = O^{15} + \gamma, O^{15} \rightarrow N^{15} + H = C^{12} + He^4.$$

As C^{12} is reproduced in almost all cases, the abundance of the catalysts is unaffected. Apparently, both reactions are equally probable at a temperature of 16×10^6 degrees which is about the central temperature of the sun. Bethe, in fact, clarifies that at lower temperatures the first reaction predominates while in the case of stars with higher central temperatures, the reaction (2) dominates.

During all these processes lot of heat energy get liberated and this heat energy comes out as radiation. What next? At this stage, the original molecular cloud would have shrunk to a very small volume because of gravity which now gets opposed by the radiation pressure along with the gas pressure and attains almost a stable equilibrium and emits radiation in the optical frequency and becomes visible! A **Star** is said to be born!!

In fact the basic physics of energy production in stars was discussed apart from Bethe by several others like C von Weizsacker, Gamow and Teller. However, the most important work for understanding the energy production in stars is due to Fred Hoyle and collaborators, M and G Burbidge, and W Fowler. They wrote a very comprehensive review article on the topic covering their work of the period 1949–1956, in 1957 which is famously known as Burbidge, Burbidge, Fowler and Hoyle (1957). A very short summary of this may be found in Narlikar (1977) under the heading 'Star as a thermonuclear reactor'. The (B^2FH 1957) paper has been considered as the most important landmark for understanding the topic of energy production in stars. Though the details of the required nuclear reactions were worked out mostly by the above-named physicists, one cannot forget to mention Sir Arthur Eddington who in 1920

(Eddington, 1920) had authored the paper 'Internal constitution of stars', wherein he sums up the possibility of hydrogen transmutation to helium in the core of the star as the main source of energy as that could have been the only source known to occur. This he argued as seen from Aston's experiments which showed that the mass of the helium atom is less than that of four hydrogen atoms, and the mass loss during the fusion of Hydrogen to Helium should be converted to energy as expressed by Einstein's formula $E = mc^2$. The released energy comes in the form of radiation that creates radiation pressure as given by $p_r = \frac{1}{3}\sigma T^4$, where σ is the Stefan's constant and T the temperature. This along with the gas pressure p_g balances the gravitational force as given by $\left(-\frac{GM(r)}{r^2}\right)$ and keeps the configuration in radiative equilibrium, governed by the equation:

$$\frac{d}{dr}(p_g + p_r) = -\frac{GM(r)}{r^2}\rho \tag{1.1}$$

ρ being the density. From the kinetic theory of gases, one knows that in a volume of gas, the distribution of molecules depends upon the temperature of the gas and for higher temperatures, the molecules are far more separated than at a cooler temperature. This obviously relates the temperature to the density of the gas and consequently to the pressure. Thus given any distribution of fluid material the pressure, density and temperature are related and this relation is called the equation of state expressed as $f(T, P, \rho) = 0$. As one will see later, this is a very important concept that plays a fundamental role in the discussion of the structure and status of any material distribution, and particularly so for the evolution of stars. The main source of energy being at the center of the star where both density and temperature are very high, the radiation from the center traverses through the stellar interior to the surface by the mechanism of radiative transfer. Without going into the details of this transport and absorption involving multiple scatterings of the photons by the intervening medium, it may be summarized that by the time the radiation reaches the surface, the temperature would have dropped down to lower values. Also, the time involved in this process of transfer seems to be very large as worked out by Eddington in the 1920s. As it is blackbody radiation the surface of the star will exhibit the colour as radiated at that temperature. For example, the sun which exhibits maximum emission in UV is likely to have a surface temperature of the order 5800°K. As stars of different temperatures radiate in different frequencies with different intensities, there exists a great variety of stellar spectra, having different characteristics relating their masses, luminosities and colour.

Stars are grouped into different categories in order of descending temperature with the names "O, B, A, F, G, K, M S". Spectral types O and B are considered to be of early type, while those with types G, K, M and S are said to be of late type. [An easy way to remember the spectral types is through the sentence, 'Oh, Be A Fine Girl/Guy and Kiss Me Sweetheart']. The brightness of stars is measured in terms of 'magnitudes' on a scale where a 6th magnitude star is exactly 100 times fainter than a 1st magnitude star. This gives a magnitude ratio of about 2.512, such that the larger the magnitude, the fainter is the star. What one sees in the sky is referred to as the apparent magnitude, while the 'absolute magnitude' or the intrinsic brightness, also called the 'luminosity', is defined as the apparent magnitude the star would exhibit, if viewed from a distance of 32.6 lys. (1 ly is the distance light would travel in a year at the speed 300,000 km/sec). The brightest visible star in the sky, **Sirius** is of magnitude −1.47, while the Sun's absolute magnitude is 4.8.

To classify the stars, Hertzsprung in 1911 and Russel in 1913 had plotted the luminosities of stars against their spectral classes indicating the colour (a way of denoting their surface temperatures). These investigations gave important information regarding the relation between the temperature and luminosity of stars. This indeed became a standard way of identifying the nature of a star from its position in the diagram popularly known as the Hertzprung–Russel (H–R) diagram. The plot depicts the Luminosity against the temperature in a way slightly different from the normal plots. The temperature decreases as one goes to the right on the abscissa, and the luminosity is plotted on the ordinate increasing upwards. A typical H–R diagram is shown in the adjoining Fig. (1.11).

It may be noticed that the stars cluster into certain parts of the diagram but yet a great majority are aligned along a narrow sequence running from the upper left to the lower right and this band is well known as the 'main sequence'. One can easily see that the hotter stars are more luminous than

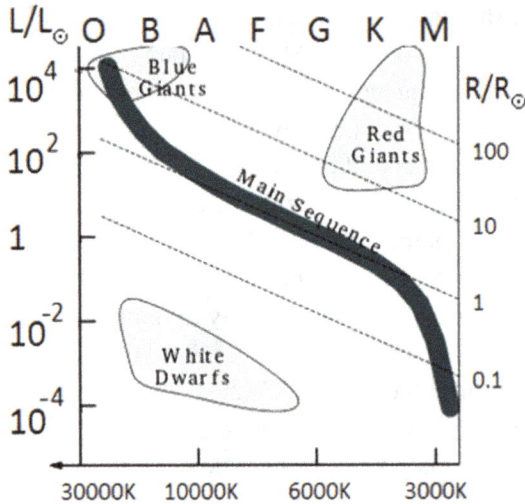

Fig. 1.11.   H-R diagram. *Image Credit: Encyclopedia Britannica*, 8th June 2021. https//www.britannica.com/science/Hertzprung–Russel diagram. Accessed June 2022.

the cooler ones as could be generally expected. As marked on the graph, one sees that there are a number of stars above the main sequence which are high in luminosity but of lower temperature. One may ask how could this be. These are indeed referred to as giants and super giants as their sizes are so enormous that their total integrated energy output is large and thus are more luminous. Similarly one finds the opposite too, a group which is of higher temperature but lower luminosity indicating that their total surface area is smaller and consequently of lower dimensions and thus are called the dwarfs.

As the brightness measured is normally expressed over a range of frequencies or wavelengths, it becomes necessary to introduce colour which for a blackbody is proportional to the temperature of the body (directly to frequency and inversely to wavelength), according to the relation $T/v_m = T\lambda_m = $ constant, v_m being the frequency and λ_m the wavelength at which the maximum emission occurs. For example, the solar emission is maximum in the UV frequency ($\sim$3700 $\mathring{A}$) and thus sun's surface temperature is estimated to be about 5800 K. This tells us that the external appearance of a star in a sense tells approximately about its location in the

H–R diagram and therefore the concept of colour-index is also used. Though this may not hold good for the entire stellar population, it appears to be so, for nearby stars and particularly for those in the Milky way galaxy.

The main sequence of the H–R diagram seems to bifurcate into two regions with the upper main sequence populated with heavy mass stars ($M > 2M_{sun}$) having convective core where energy production is mainly through the CNO cycle. The lower main sequence has stars with mass less than about 1.5 M_{sun} and their cores are mostly radiative and the energy production is mainly from the p-p reaction. In fact, the H–R diagram represents an evolutionary sequence as its population is from different compositions, temperatures and ages. Massive stars are hottest in their central region and radiate at a greater rate converting mass into radiant energy and thus spend much less time on the main sequence as compared to the sun. When all the hydrogen in the star's core is used up there would be changes in its temperature and surface brightness. As the core becomes less rich in helium towards the center, the star moves off the main sequence as the overall equilibrium gets disturbed due to the discontinuous change in the composition. At this stage in order to readjust itself the outer portion of the star expands to a very large size while the inner core contracts due to gravity. The outer region becomes cooler due to expansion and appears red in colour. Thus one has 'Red Giant' stars, which are normally about 100 times bigger than their size when on the main sequence.

It is necessary to realise that all collapsing clouds may not end up as stars. Sometimes the gas and dust pressure inside the core may be just sufficient to stop collapse and keep gravity balanced. Such configurations are called 'protostars', and their temperature is not high enough to radiate in the optical frequency. Such cores could emit in the microwave and infrared wavelengths which could last for about a million years. Evolution of protostars has been discussed by Hayashi *et al.* (1966). For a detailed discussion of the thermal behavior and general features of evolution of a protostar of about one solar mass over a large range of temperature and density, one could refer to (Hayashi *et al.*, 1965). Protostras are generally very young, still gathering mass from its parent molecular cloud which for a low mass star lasts about half a million years. This phase begins when molecular cloud fragments collapse under self-gravity to an opaque pressure-supported core and with the decrease of infalling gas and dust finally lead to a pre-main sequence star. Within its deep interior, the temperature is lower than that of a normal star. Though the protostars generate energy which comes from radiation liberated at the shocks on its surface and on its surrounding disks. The radiation thus liberated needs to travel through interstellar dust which absorbs and reradiates at lower frequencies than optical and could thus be observable at IR and FIR wavelengths.

In Chapter 2, we take up the narration of the evolution of stars off the main sequence, particularly giants and supergiants.

Chapter 2

Stellar Evolution Beyond the Red Giants and Super Giants

2.1 Red Giants and Planetary Nebulae

How do energetics work at the red giant phase of stellar evolution? As the hydrogen at the center is depleted the burning hydrogen in the shell is the main source of energy. This outward-moving hydrogen getting converted to helium, slowly increases the mass of the helium core. The size-reduced core with increasing mass of helium gets heated to such a temperature that the conversion of helium to carbon happens in such rapidity, it is called a 'helium flash'. The core temperature at this stage is about 10^6 K and the density is about 10^8 kg/m^3 which would be sufficient to cross the coulomb barrier for helium fusion. Three helium nuclei combine to form an excited state of carbon which then decays to reach the normal state. The energy of the excited carbon that first forms is only slightly higher than the combined energy of Berilium^8 nucleus and an α particle. In the helium-burning core, further nuclear reactions continue with fusion of helium with newly formed nuclei resulting in the formation of higher elements of the periodic table, O^{16}, Ne^{20}, Mg^{24}, Si^{28}, S^{32}, etc. The atomic number increasing by units of 4 indicates the addition of an α particle for every new element and thus this part of nucleosynthesis is called the α-process. Some details of these reactions with associated energetics may be found in Lang (1974) who mentions that the reaction products of the C N O cycle might also produce neutrons through this process and these neutrons are captured in later reactions for furthering the element formation.

How long can this continue? The Coulomb repulsion of nuclear charges which was stopped earlier will now return and as such the synthesis of higher elements need some other processes. At the conditions of helium burning the star would have predominantly carbon C^{12} and oxygen O^{16} nuclei, and if the temperature is greater than about 800 million $°K$, Carbon nuclei react with themselves producing Magnesium[23] which with the addition of a neutron turns into (O^{16} + 2 He^4). With the increase of temperature to about a billion (10^9) K, Oxygen nuclei combine to form Magnesium[24] + 2 helium nuclei He^4. As one reaches the end of carbon and oxygen burning, the star would have sulphur and silicon as the most predominant nuclei along with magnesium, and at this stage, silicon transmutes to Nickel[56] which decays to Cobalt Co^{56} and Iron Fe^{56}. With the arrival of the iron group, this chain reaction stops as these elements are most stable. In fact, these heavier elements seem to form by the process of 'neutron capture' reactions which continue on to form new elements heavier than iron group elements. The processes are termed as s and r processes as some reactions will have neutron capture 'slower (s)' than the beta-decay ($n \rightarrow p + e^- + \bar{v}_e$) life time. If the flux of neutrons is very strong then this capture can occur faster than beta decay and thus is called the 'rapid (r)' process. [B^2FH]. It is believed that while the s-process yields more of proton-rich nuclei, the r-process yields more of neutron-rich nuclei. Also, it has been suggested that the s-process may hold in red giant type stars, where the original dust and gas cloud was formed from the galactic material which contained elements like hydrogen, carbon, oxygen, neon, magnesium and some iron group elements whereas the r-process could occur in explosive environment of a supernova (Nanlikar, 1977).

The following picture is the first direct image of a red giant star taken with NASA's Hubble Space Telescope. It is called Alpha Orionis, or Betelgeuse. It is in fact a red supergiant star sitting at the shoulder of the winter constellation Orion the Hunter (Fig. 2.1).

This Hubble image reveals a huge ultraviolet atmosphere with a mysterious hot spot on the stellar surface. The enormous bright spot, twice the diameter of the Earth's orbit, is at least 2,000 degrees Kelvin hotter than

Fig. 2.1. Atmosphere of Betelgeuse — Alpha Orionis. Hubble Space Telescope — Faint Object Camera, *January 15, 1996; A. Dupree (CfA), NASA, ESA.*

the surface of the star. Betelgeuse is so huge, that if it replaced the Sun at the center of our Solar System, its outer atmosphere would extend past the orbit of Jupiter (scale at lower left).

As nucleosynthesis continues in the shrinking core, the burning helium shell and the hydrogen shell expand upwards but get unstable due to inhomogeneity. At the same time in the case of low mass stars, all the elemental formation in the core would have ended with the formation of iron group elements and thus the core becomes cold. Gravity again takes over and the gravitationally collapsing core produces shock waves that ripple the outer layers and exert pressure on the outer shells of gas and dust. This outer shell gets ripped off the star as an expanding wispy shell which is called a **planetary nebula**.

The nebulosity gets its name from the fact that from the earth it appears like a planet.

Some other famous examples of planetary nebulae are the following: The Ring nebula is in the constellation Lyra. It was discovered by Charles Messier while searching for comets (Fig. 2.3).

Planetary Neb NGC 2610
R:G:B=[N II] 400s:[O III] 400s:He II 400s
KPNO 2.1m, Ref: Balick 1987 AJ 94 671

Fig. 2.2.

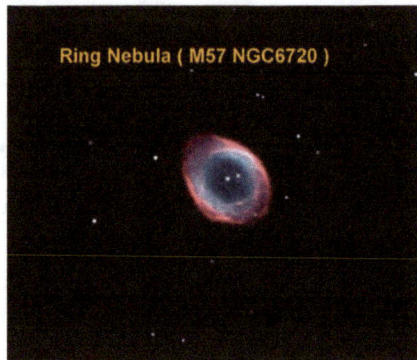

Ring Nebula (M57 NGC6720)

Fig. 2.3.

The Dumbbell Nebula is in the constellation Vulpecula, at a distance of about 1,360 light-years. It was the first such nebula to be discovered, by Charles Messier (Fig. 2.4).

The Boomerang Nebula is in the constellation of Centaurus, 5000 light-years from Earth. In 1995, using the 15-metre Swedish ESO Submillimetre Telescope in Chile, astronomers revealed that it could be the coldest place in the universe found so far with a temperature of around −272°C or 1K (Fig. 2.5)

Fig. 2.4.

Fig. 2.5.

The Boomerang Nebula is in the constellation of Centaurus, 5000 light-years from Earth. In 1995, using the 15-metre Swedish ESO Submillimetre Telescope in Chile, astronomers revealed that it could be the coldest place in the universe found so far with a temperature of around $-272°C$ or 1K (Fig. 2.5).

The Cat's Eye Nebula is in the northern constellation of Draco, discovered by William Herschel (Fig. 2.6). Planetary nebulae are supposed to be important sources for the presence of gas in the interstellar medium. They are mostly small, being about two to three lys across with an average mass around 0.3 M_{sun} and are denser than most H II regions but are much more regular than H II regions. Helix nebula also known as NGC 7293 in the constellation Aquarius is considered as one of the largest planetary

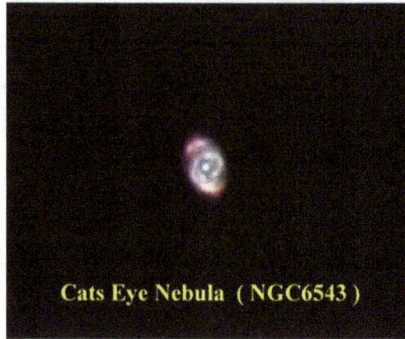

Fig. 2.6.

Fig. 2.7.   Spitzer IR.

Note: IR image of Helix nebula and Visual image of Helix.

nebulae. (IR image of Helix nebula and its Visual image) Helix nebula also known as NGC 7293 in the constellation Aquarius, is considered to be one of the largest planetary nebulae.

High-resolution images of planetary nebula usually reveal tiny knots and filaments and their spectrum seems to be basically the same as that of the H II region containing bright lines from hydrogen and helium. The spectra of planetary nebulae seem to reveal that they are expanding from their central star (the core of the red giant) at an average speed of about 50 km/sec.

It is surmised that the gravitational pull of the core on the expanding shell is rather small being proportional to the distance in between which

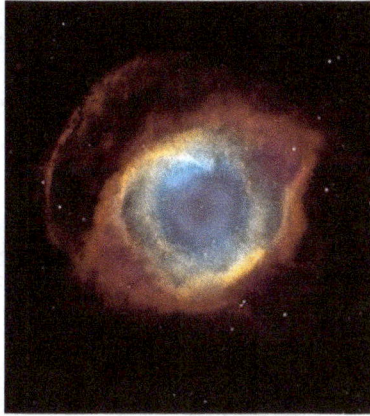

Fig. 2.8. Hubble Optical.

Note: IR image of Helix nebula and Visual image of Helix.

appears to be consistent with the idea that the entire mass of gas must have been ejected in a brief time due to some instability. This expanding gas shell soon gets mixed up with the interstellar gas as discussed by J.S. Mathis (2019). Nova T Pyxidis is a recurrent nova the ejecta of which consists of some 2,000 gaseous blobs in a volume one light-year across. The high-resolution HST image of the Nova T Pyxidis demonstrates that the ejecta of most novae are expanding, gaseous shells. (Fig. 2.9).

2.2 Contracting Core and White Dwarfs
Equation of state

Before going on to the discussion of the contracting core, let us take a look at the equilibrium structure of fluid configurations which all stellar interiors are made of. As Eddington points out (Eddington,1957) J.H. Lane in 1870 was the pioneer to investigate the temperature distribution within a star followed by others like Lord Kelvin, J. Ritter and finally by R. Emden. Lane's result was that if a star contracts the internal temperature rises so long as the material is sufficiently diffuse to behave like a perfect gas. It sets forth the idea that the change of temperature is necessary to maintain equilibrium. He asserts that, in principle, the ideas of Hertzsprung and Russel (H–R diagram) were a revival of those

Fig. 2.9.   T Pyxidis. (Hubble telescope picture of T Pyxidis, from a compilation of data taken on 26, Feb 1994, and June 16, Oct. 7, and 10 Nov 1995, by the Wide Field and Planetary Camera 2. Mike Shara, Bob Williams, and David Zurek). (Space Telescope Science Institute); Roberto Gilmozzi (European Southern Observatory); Dina Prialnik (Tel Aviv University); and NASA, Public domain, via Wikimedia Commons.

of Lane and Lockyer, with the novel point of adopting discussion to observational data. To quote, *"the stars start to be visible as cool red stars of type M with low density and enormous bulk. They contract and in obedience to Lane's condition rise in temperature passing up the spiral series K, G, F, to A and B-i.e. the reverse of the previously accepted order. At some stage of the contraction the density becomes too great for the perfect gas laws to apply, the rise of temperature is checked and ultimately the star cools down and returns down the spectral series and ends in extinction."*

As Chandrasekhar has emphasised, the fundamental problem is the study of equilibrium configurations in which the pressure P and density ρ are connected by a relation of the kind

$$P = K\rho^{\frac{(n+1)}{n}}, \tag{2.1}$$

where K and *n* are constants. Such configurations are termed as *polytropes of index n*. The equations that govern such fluid distributions are along with (2.1) having $(P = p_g + p_r)$ two more as given by

$$\frac{dM(r)}{dr} = 4\pi\rho r^2 \tag{2.2}$$

and $$\frac{1}{r^2}\frac{d}{dr}\left(\frac{r^2}{\rho}\frac{dp}{dr}\right) = -4\pi G\rho. \tag{2.3}$$

Among different types of equations of state one considers, the adiabatic equation of state as given by $PV^\gamma = \text{const}$, $TV^{\gamma-1} = \text{const}$, $P^{\gamma-1}T^\gamma = \text{const}$, for polytropes are the most discussed. Here, $\gamma = \left(1 - \frac{1}{n}\right)$ represents the ratio of specific heat at constant pressure c_p and the specific heat at constant volume c_v, generally referred to as the adiabatic index. As has been discussed in most standard texts, one introduces the Emden variables θ and ξ related through the quantities $\rho = \lambda\theta^n$ and $r = \alpha\xi$, where $\alpha = \left(\frac{(n+1)K}{4\pi G}\lambda^{\left(\frac{1}{n}-1\right)}\right)^{\frac{1}{2}}$ and using the equation of state $P = K\lambda^{\left(1+\frac{1}{n}\right)}\theta^{(n+1)}$, the equation of equilibrium, is written the form:

$$\frac{1}{\xi^2}\frac{d}{d\xi}\left(\xi^2\frac{d\theta}{d\xi}\right) = -\theta^n \tag{2.4}$$

which is the well-known *Lane–Emden equation of index n*. [those interested in getting the detailed mathematics and the discussion of its applications may refer to Chandrasekhar (1939); Zeldovic and Novikov (1971)]. If $\lambda = \rho_c$, the central density and the boundary conditions for solving the LE equation are $\theta = 1$ at the centre ($\xi = 0$) and $d\theta/d\xi = 0$ (so that there is no cusp at the centre), one can solve the LE equation for integral values of $n = 0, 1$, or 5, while for other values one solves it numerically. However, as one could expect that for a particular value of the central density, one would have a particular combination of the radius R and mass M.

2.3 Chandrasekhar and the Story of Mass Limit

What happens to the contracting core? As thermonuclear fusion had stopped in the core, there was no source of heat after the formation of iron group nucleii. The cold matter gets compressed by gravity to higher density to such an extent that for understanding the structure of matter classical hydrostatics, or statistical mechanics are no longer suitable as one has to treat the matter quantum mechanically, in which context it would be useful to recall some basic aspects of atomic physics which are explained only by quantum mechanics. It all started with the experiments of Rutherford, who successfully showed through experiments of scattering of alpha particles by a thin gold foil that most of the matter in atoms could be confined to their nucleus and electrons could be moving around them. The coulomb interaction between the central nucleus with positive charge and electrons with negative charge could be keeping the atom in Coulomb equilibrium. However, there were difficulties in explaining the radiation emitted, as electrons in classical orbits would get accelerated and radiate continuously losing energy and falling into the nucleus. Apart from this observation, there were also problems in explaining the presence of sharp, discrete spectral lines which indeed were the characteristics of the emitting atom. As recognized, this was indeed a complete breakdown of the classical theory of radiation as emphasised by Niels Bohr who further initiated a new explanation using the notion of quantum (photon) suggested by Planck and identified by Einstein through the photoelectric effect. Bohr's theory was that the system of electrons and the nucleus which constitute an atom can exist only in certain special states characterised by discrete values of total energy called the stationary states wherein the atom can remain indefinitely without radiating.

Then where does the emission come from? For that, he explained that only when an electron jumps from one state with energy E_i to another state with energy $E_f < E_i$, the loss in energy gets emitted as radiation such that the frequency of the radiation follows the relation

$$E_i - E_f = h\,\nu. \tag{2.5}$$

The discrete energy values of stationary states are called the energy levels. While doing so Bohr adopted the views of both Planck (discreteness of energy) and Einstein relating energy and frequency. Bohr's main contention was that the special orbits of the electrons in an atom, the stationary states are those in which the angular momentum of the electron about the orbital center normally denoted by l, is an integral multiple of $\hbar$ which is equal to $h/2\pi$. Thus for a circular orbit, one has $mva = n\hbar$, which is known as Bohr's quantisation condition. Considering the atom as a system where a charged particle (electron) is moving around a center in a circle of radius r the electrostatic attraction force e^2/r^2 is countered by the centrifugal force mv^2 and thus one can find the radius r to be

$$r = (n\hbar)/me^2. \tag{2.6}$$

As an example when these notions are applied to the hydrogen atom, one can work out the quantised energy levels to be given by the formula

$$E_n = -me^4/2n^2h^2, \quad n = 1,2,... \tag{2.7}$$

$n = 1$ corresponds to the state of lowest energy which is also called the ground state. Having thus seen that electrons in atoms will have different orbitals depending upon their energies, one could ask, can more than one electron be present in the same orbit? This according to Pauli is not possible unless they have opposite spins. This is known as *Pauli's exclusion principle* according to which 'no two fermions (particles with half-integer spin-like electrons, protons) can occupy the same state as they both cannot have the same quantum numbers'. This brings in the question of what exactly one means by saying the same quantum numbers. What if their momenta or the positions differ infinitesimally? Then comes another important aspect of quantum mechanics, *Heisenberg uncertainty principle* which restricts the possibility through the following relation: if δp is the difference in their momenta and δr the difference in their positions then one should always have $\delta p \delta r \approx \hbar$. Consequently, there is always an uncertainty in the simultaneous measurement of momentum and position, as given by $\Delta p_x \Delta x \sim \hbar$. With these fundamental restrictions, it appears that each particle is characterised by its position in a six-dimensional frame work (cell) of coordinate space and components of momentum called the *phase space* whose volume is given by $\delta x \delta y \delta z \delta p_x \delta p_y \delta p_z = \hbar^3$. In order to understand as to the number of electrons that can be held in a given cell consider the radius and the energy of the particle in its lowest orbit ($n = 1$) as given in (2.6 and 2.7). If a single electron is orbiting the nucleus with charge Ze then one replaces e with Ze in the formulae which means that with increasing atomic number Z the radius of the ground state will decrease whereas the absolute value of its energy will increase as Z^2. If instead of one electron Z number of electrons are there, the atoms forming the natural system of elements will become smaller and more tightly bound. Though there is the question of electrostatic repulsion, calculations have shown that this repulsion cannot prevent atoms of heavier elements from shrinking to a smaller size. This would enforce the atomic volumes to decrease continuously and rapidly from hydrogen to uranium. Apart from this if all the electrons of an atom were located at the ground level (lowest energy level) then it would be more and more difficult to extract an electron from light to heavy elements. However, either of these two features was completely absent in the experimental results. Particularly, the experimental studies of the ionisation potentials seem to have revealed that with the addition of more and more electrons, volumes occupied by various quantum states shrink but the number of states occupied increases keeping the external diameter of the atom remaining approximately constant. Wolfgang Pauli has suggested that this difficulty could be settled if one *permits only two electrons to occupy any given quantum state described by three quantum numbers viz., the radial, azimuthal and the orientationl ones.* Using this principle later called as *Pauli's exclusion principle*, Bohr and colleagues were able to construct satisfactory models for all known atoms of the periodic table. The main consequence of such a situation is that when matter in any system gets packed closely such that all the lowest electron states are filled, any further contraction of the system can force the electrons to assume much higher momenta normally found in tenuous gases. Such closely packed gas of fermions is said to be *degenerate* (Mathews and Venkatesan, 1976; Harwitt, 1973; Gamow, 1966).

Eddington and Fowler studied this aspect of stellar evolution where the radiative equilibrium could not sustain gravitational attraction and the star keeps on contracting and, had expressed that 'matter can exist in such a dense state ($>10^6$ gm/cc) only if it has sufficient energy such that the electrons are no longer bound to their parent nuclei and are free to escape.

As Fowler emphasises (Fowler 1926) at such temperatures and densities one needs to apply the Fermi–Dirac statistics as applied to fermions (electrons are fermions with half integral spins). The essential feature of the F–D statistics is that it is governed by the Pauli exclusion principle, in which two fermions cannot occupy the same cell of size $(h/2\pi)^3$ in the phase space, h being the Planck's constant. In other words, no two electrons with same quantum numbers can occupy the same shell. In such a situation if the particle density is increased due to the gravitational contraction, the added particles are forced to occupy states of higher energy because the lower energy states are all progressively filled resulting in degenerate matter. This builds up a pressure in the configuration of fermion gas and this pressure is known as the *degeneracy pressure*. If all the energy states below the Fermi energy are filled then it is known as fully degenerate which is also recognised as the *zero temperature fermion gas*. At this stage, the degeneracy pressure is so high that gravity can no longer push the particles further resulting in an equilibrium configuration with the electron degeneracy pressure opposing gravity. This would lead to configurations having high density and settling down as a dead star. Such configurations are of much lower luminosity though their mass could be in the range of a solar mass. The main reason for this could be the fact that the radius of such stars is very much smaller than the ones on the main sequence of the H–R diagram. The companion of the star Serius is supposed to be one such dead star, and because of their smaller radii are termed 'dwarf stars', compared to the stars on the main sequence. For a

S.Chandrasekhar, the Indian born astrophysicist, after finishing his Masters in Presidency College, moved to England on a scholarship in 1930. On the way to England by boat he worked on the problem of polytropes, but with relativistic equation of state and had obtained the limiting mass later to become known as 'Chandrasekhar limit' for white dwarfs. His final paper on this was written in 1935. Almost fifty years later in 1983 he was awarded the Nobel Prize in Physics, alongwith William Fowler an astrophysicist from Caltech.

1910-1995

Fig. 2.10.

given effective temperature they are much fainter but for the same luminosity their effective temperatures are much higher and thus they are termed as the *White dwarf* and appear at the bottom left of the H–R diagram. During the late nineteen twenties and thirties, the theory of white dwarfs was the most engaging problem of stellar structure, and several astrophysicists like Eddington, Stromgern, Anderson, Fowler and Chandrasekhar were involved in trying to solve the issue.

As mentioned earlier (Fowler) in 1926 (Eowler, 1926) was the first to suggest the possibility of such configurations, where the electron degeneracy pressure opposes gravity in setting up the equilibrium state irrespective of the core density stopping the gravitational collapse and this idea was fully supported and accepted by Eddington, Stoner and Anderson. According to Fowler's analysis, the contracting core when stopped by the degeneracy pressure remains so as a dead star and no further evolution happens irrespective of the mass and density of the star. In order to do this analysis Fowler had adopted the *non-relativistic equation of state*, $p \propto \rho^{5/3}$. It was clear that any detailed analysis of such a configuration must take into account the interplay between three fundamental properties namely the mass, the radius and the luminosity of the star. Chandrasekhar reanalysed this problem in a series of publications during 1930–1935, considering a completely degenerate electron gas core as the one in which all the lowest quantum states are occupied and ascertained that at large densities like that of a white dwarf, the electrons would be moving at *relativistic speeds (close to the velocity of light) and thus the equation of state* to be considered should be '*relativistic*'.

Chandrasekhar made a detailed analysis of the polytropic configurations with different equations of state and showed rigorously that there is a *limiting mass* for the collapsing configuration to get to an equilibrium state and any mass above this limit, makes the core collapse continuously overcoming the electron degeneracy pressure and thus not reaching the white dwarf state. Unfortunately, this issue became very controversial in those years with Eddington, declining to accept Chandrasekhar's analysis, particularly the role of *relativistic treatment of the electron gas core*. For a lucid treatment of this controversy between Eddington and Chandrasekhar one may refer to the popular books (1) *Chandra* by (Wali Kameswar, 1991) and (2) '*Chandrasekhar limit*' by (Vekatraman, 1992).

Controversies apart, the stellar evolution does seem to follow the way Chandra described and the white dwarfs do have a limiting mass of 1.4 M_{sun} which indeed is known as the *Chandrasekhar limit*.

For a detailed discussion on degenerate stellar configurations and the theory of white dwarfs, the most authoritative account may be found in the book *An Introduction to the Theory of Stellar Structure by* (Chandrasekhar, 1939).

As he says' since the radius of a white dwarf is very much smaller than that of a star on the main sequence, for a given effective temperature, the white dwarf will be much fainter than the ones on the main sequence. On similar grounds, for a given luminosity a white dwarf appears to have higher effective temperature than its counterpart on the main sequence. Emphasising this point, Eddington, with the example of the companion of Sirius in mind, says, 'strange objects which persist in showing a type of spectrum entirely out of keeping with their luminosity, may ultimately teach us more than a host which radiates according to the rule' (E A 59).

Chandrasekhar's analysis using the relativistic equation of state for the degenerate electron gas indeed answered the question by showing that the white dwarf stage is the end of individual low mass stars which may be treated as dead stars. One could say that this analysis of Chandrasekhar was the first discussion of 'Relativistic astrophysics', beyond the solar system, keeping in mind the significant application of the general theory by Einstein himself to explain the perihelion motion of Mercury and the bending of light by the sun's gravitational field.

As is known under normal conditions ionized gas recombines when it cools. However, the density in the interior of white dwarfs is sufficiently high that the gas remains ionized due to pressure ionization. Also due to their strong degeneracy ($E_F \gg kT$) they do not contribute appreciably to heat capacity. With the electron thermal conductivity being very high, it keeps the interior isothermal. Larger the mass of the white dwarf smaller is its radius. For a solar mass white dwarf the radius is about one hundred times smaller than the solar radius. The Chandrasekhar limit above which no stable white dwarf configuration can exist as only infinite central pressure can keep the star from further collapse. One can summarize the scenario as given by (Harwitt, 1982).

The main reason for the limit could be as follows: The pressures in the cases of relativistic and non-relativistic equations of state being $P \propto \rho^{4/3}$, $P \propto \rho^{5/3}$, one finds for their gradients

$$\frac{dP}{dr} \propto \rho^{\frac{1}{2}} \frac{d\rho}{dr}, \quad \text{and} \quad \frac{dP}{dr} \propto \rho^{\frac{2}{3}} \frac{d\rho}{dr}. \tag{2.8}$$

As the gravitational pressure gradient in either case is given by

$$\frac{dP}{dr} \propto -\frac{\rho(r)}{r^2} \int_0^r 4\pi r'^2 dr', \tag{2.9}$$

in a very crude approximation using $\rho \propto M/R^3$, One can get the conditions as given by

$$dP/dr \propto M^{5/3}/R^6 \text{ (non relativistically)}, \tag{2.10}$$

$$dP/dr \propto M^{4/3}/R^5 \text{ (relativistically)}, \tag{2.11}$$

$$\text{and } dP/dr \propto M^2/R^5 \text{ (gravitationally)}. \tag{2.12}$$

Referring to the central density of the star as ρ_c one finds $R \propto \rho_c^{(1-n/2n)}$, $M \propto \rho_c^{(3-n/2n)}$, which for n = 3/2 gives $\propto \rho_c^{-1/6}$, and $M \propto \rho_c^{+1/2}$. Combining the two one has $R \propto M^{-1/3}$.

As can be noted the dependence on the radius is the same for the gravitational and the relativistic case whereas for the non-relativistic case, it is lower. This shows as the radius decreases the pressure gradient increases at the same rate as the gravitational compression. For smaller masses, the central pressure is determined more by the non-relativistic approximation which lends to a possible equilibrium stage. On the other hand for more massive objects the central density becomes so high during contraction reaching the relativistic regime and further contraction no longer leads to equilibrium. The Chandrasekhar limit where the central density is almost infinite thus becomes a transition from nonrelativistic to relativistic regime for a contracting core of degenerate electrons reaching a maximum mass for stable equilibrium.

A view of the globular cluster M4 (fourth object in the Messier catalog of star clusters) and (nebulae) nearest globular cluster to Earth (about

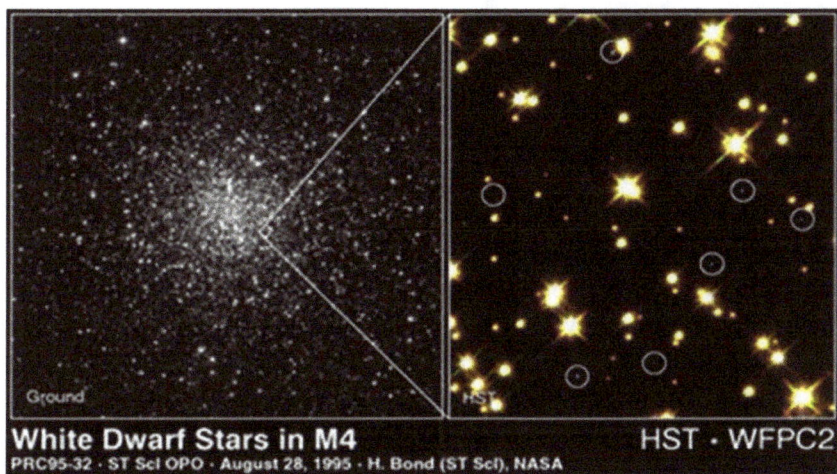

Fig. 2.11. *Image Credit*: Kitt Peak National Observatory 0.9 meter telescope M.Bolte, Uni of California, Santa Cruz.

7,000 light-years away) seems to contain more than 100,000 stars. This cluster was the target of a Hubble Space Telescope search for white dwarf stars. In this view from a ground-based telescope, it appears that a large number of red giant stars are predominant. The field is 47 light-years across. The box (right of center) in the left panel shows the small area that the Hubble telescope probed.

In the right panel, Hubble reveals a total of 75 white dwarfs in one small area within M4, out of the total of about 40,000 white dwarfs that the cluster is predicted to contain. The Hubble results will allow astronomers to refine theoretical predictions of the rate at which white dwarfs cool, which in fact is an important prerequisite for making reliable estimates for the age of the universe and of the Milky Way galaxy, based on white dwarf temperatures.

As mentioned earlier, if the white dwarf is in a binary system then it accretes matter from its companion or the surrounding interstellar matter and thus naturally increases the gravitational instability leading the star to explode as a nova or a supernova depending upon the amount of mass accumulated. As mentioned, if one of the binary components evolve much faster than the other and end up as a white dwarf, the mass transfer between the primary and the secondary can lead to nova outburst. This feature seems to exist in close binaries with short periods of the order ~4 hrs. The inflowing mass having high angular momentum forms an accretion disc around the white dwarf, with viscous forces helping the accretion. The light curves of such novae show a rapid rise initially followed by a slow decay. The time interval between the maximum brightness and reduction by about 3 to 4 magnitudes occurs over a period of few weeks later followed by gradual decline over a period lasting up to ten years. Before reaching the maximum the nova exhibits broad absorption lines that change to A or F supergiant type and shortly after, the maximum exhibits strong symmetrically broadened emission lines along with blue-shifted absorption lines. An important feature of these novae is that at their maximum, with luminosities around ten to hundred times the solar luminosity, the decline in brightness is more rapid than the ones with lower luminosity. This allows one to use them as distance indicators, even at large distances where their changes in angular diameters are difficult to measure.

Chapter 3

Star Clusters

In our *Milky Way galaxy,* one can see groups of stars being held together by their mutual attraction, forming spherical clusters like the one shown in Fig. 3.1.

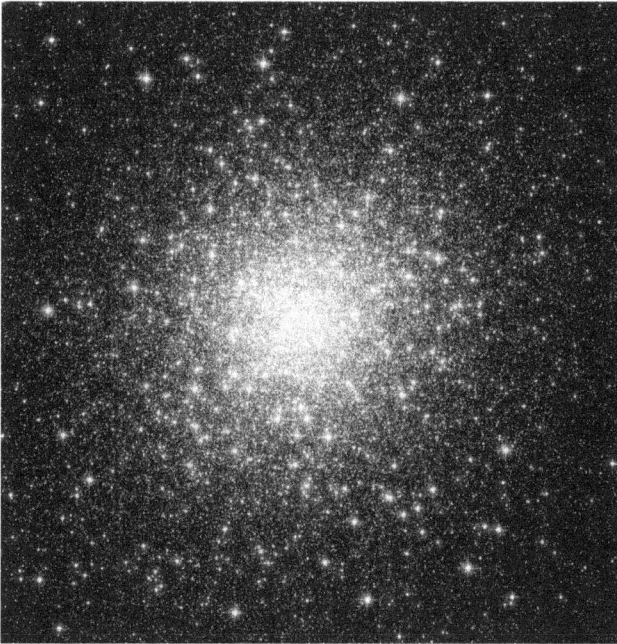

Fig. 3.1. M2: The first globular cluster to be added to the Messeir series and is called ω Centauri. *Credits*: NASA, ESA, STScI and A. Sarajedini (University of Florida).

These globular clusters generally contain a large number of stars, anything going from hundreds of thousands to about millions as seen from M2 which has about 300,000 stars. As the diameter of a cluster is about 100 parsecs, the density would range much higher than in the solar neighbourhood with a larger percentage in the centre of the cluster. In contrast, there are open clusters which are less populated (from about 100 to 1000 stars) with diameters of about few to couple of dozen parsecs, but still higher than the density in the solar neighbourhood. Galaxies have a large number of globular clusters and the much debated issue is whether the galaxies are born out of clusters or the clusters are formed within the galaxy. It appears that there are about 150–200 globular clusters in our galaxy (MilkyWay) which are distributed spherically over a radius of 100 kpc with most of them concentrated near the centre of the galaxy constellation, Saggitarius. It appears that the stellar population in globular clusters are of higher age as most of their bright members have moved away as seen in the H–R diagram. The Messier (M) and the new general catalogues (NGC) of nebulae, clusters, clouds and galaxies seem to be the initial attempts to catagorise these celestial objects. Our neighbouring galaxy 'Andromeda' is referred to as M31 as also NGC 224. It is observed that younger clusters have more stars than their older colleagues, as in the older ones the disintegrating factors are several. While encounters between stars may increase the escape velocity for some due to excess energy, the tidal forces from within the cluster may have shunted some out. Sometimes, encounter with massive interstellar clouds could also have triggered the expulsion of stars from the boundary of the clusters. While open clusters may last for about a billion to hundred million years, globular clusters survive for about a hundred billion years.

Planetary nebulae are supposed to be important sources of gas in the interstellar medium. They are mostly small, being about two to three lys across with an average mass of around 0.3 M_{sun} and are denser than most H II regions but are much more regular than H II regions. High-resolution images of planetary nebula usually reveal tiny knots and filaments and their spectrum seems to be basically the same as that of the H II region containing bright lines from hydrogen and helium. The spectra of planetary nebulae seem to reveal that they are expanding from their central star (core of the red giant) at an average speed of about 50 km/sec.

Consequently, it is surmised that the gravitational pull of the core on the expanding shell is rather small being proportional to the distance in between, which appears to be consistent with the idea that the entire mass of gas must have been ejected in a brief time due to some instability. This expanding gas shell soon gets mixed up with the interstellar gas, as discussed by J.S. Mathis (2019). The expanding gas shell which is due either to radiatively driven or centrifugally driven (due to high rotation of the central star) winds can be one of the ways of mass loss from the star and this seems to be particularly more in the red giant phase of the star. In a limited way, the solar wind from the Sun could be an example of thermally driven wind due to the high temperature of the solar carona (Parker, 1958). Planetary nebulae seem to depict this phenomena of mass loss from the star which as mentioned is the result of the blowing off of the outer shell of the star. Another important factor in the lives of some massive stars for mass loss is Supernovae which in fact are of two types with different characterisations. Before taking up this topic, it is useful to consider another important feature of stellar configurations: the **binary system of stars**.

It appears that a large part of the stellar population do appear as a system of **two stars: the binary**. The stars are located such that they are rotating around a common centre of mass which in that frame make the stars appear stationary. According to the definition of the gravitational potential, every lump of mass will have at a point outside the potential GM/r, depending upon the distance r from the body's mass centre and the mass M enclosed in a spherical volume with r as the radius. Since there could always be heterogeneity in the distribution of matter in the enclosed volume, the equi-potential surfaces need not be spherical. Thus, in the case of a binary star, the equipotential surfaces around the two masses will be as shown in Fig. 3.2. In any binary system, depending upon the individual masses, the rate of evolution will be different, and thus, one of them may reach the end stage to become a compact star like a white dwarf as shown in Fig. 3.3.

As may be seen from Fig.3.2, there are five significant points marked L1–L5. They are called the Lagrange points and if one of the stars, say, M2 evolve faster and reach the size of a compact object (like a white dwarf), and the second one M1 evolve slower and getting to giant size

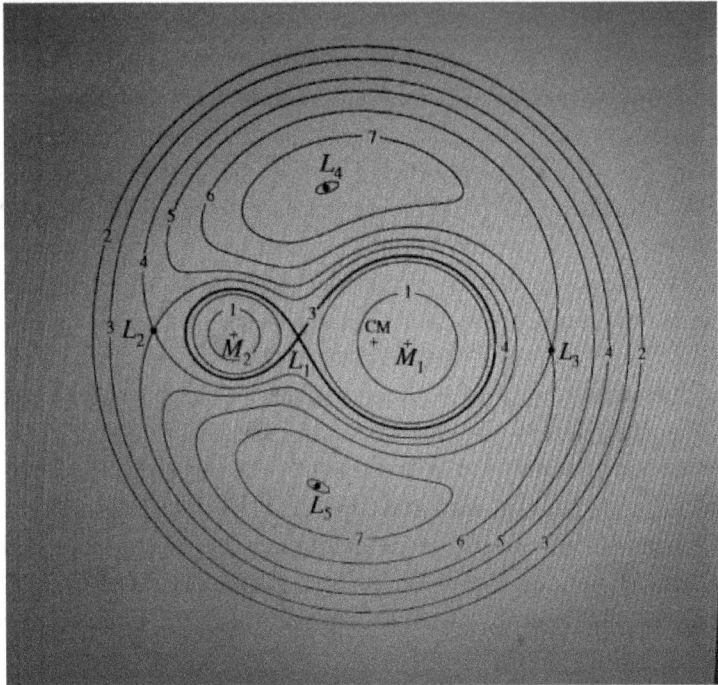

Fig. 3.2.

Credits: Adopted from Frank et al 2002 (figs 4.3 and 4,4)

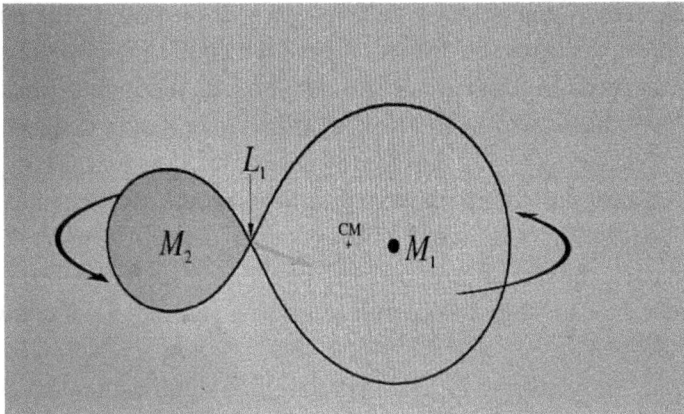

Fig. 3.3

Credits: Adopted from Frank et al 2002 (figs 4.3 and 4,4)

swells up and fills the region surrounding the lobe, called the 'Roche lobe', as shown with the intersecting point L1, the filled mass then gets pulled in by the compact object, a process termed as Roche lobe overflow (source of line drawings of Frank *et al.*, 2002). In such a situation, the white dwarf accreting mass from its companion gets heavier and thus overcome the Chandrasekhar limit which destabilises it and the star evolves further. Depending upon the amount of matter deposited, the compact object explodes in a nova or a supernova. This feature seems to exist in close binaries with short periods of the order of ~4 h. The inflowing mass having high angular momentum forms an accretion disc around the white dwarf, with viscous forces helping the accretion. The light curves of such novae show a rapid rise initially followed by a slow decay. The time interval between the maximum brightness and reduction by about three to four magnitudes occurs over a period of few weeks later followed by gradual decline over a period lasting up to 10 years. Before reaching the maximum, the nova exhibits broad absorption lines that change to A or F supergiant type and shortly after, the maximum exhibits strong symmetrically broadened emission lines along with blue-shifted absorption lines. An important feature of these novae is that at their maximum, with luminosities around 10 to 100 times the solar luminosity, they decline in brightness more rapidly than the ones with lower luminosity. This allows one to use them as distance indicators even at large distances where their changes in angular diameters are difficult to measure. In the case of binary with the compact object being a white dwarf, as the accreting matter falls in, it gets ignited and shines brightly on the white dwarf for a while. For observers from earth, it looks as though a new star has appeared and thus the phenomenon was called a **Nova**, meaning new. This feature seems to be quite common though it appears that all classical novae are not alike. Generally, this phenomena occurs in the initial stage of nucleosynthesis with hydrogen burning to form helium and when this process happens rapidly (runaway process), explosion could occur. When the process of accretion dumps in a large amount of matter onto the companion white dwarf increasing its mass and allowing gravity to take over but the process of nucleosynthesis continues on forming higher elements, this would make the white dwarf collapse further like cores of mass <u>higher</u> than 1.5 M_{sun},. In such configurations, gravity compresses the matter further as electron degeneracy pressure would not be sufficient to halt the collapse as shown by Chandrasekhar. At this stage of evolution, the star depending upon its mass will have the core consisting of positive nuclei (helium in lighter stars and iron in the heavy ones) surrounded by relativistic gas of electrons. As the white dwarf cools, the core will consist of predominantly Iron-56 with small amounts of cobalt and nickel towards the centre and lighter nuclei like chromium and manganese towards the outer regions. As the central density increases (to around 10^{10} gs/cc), for a given composition, the electron Fermi energy increases to the point where the electrons are driven into the nucleus and through the process of *inverse beta decay* form neutrons liberating neutrinos:

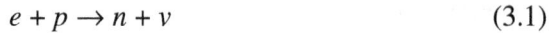

$$e + p \rightarrow n + v \tag{3.1}$$

As the electron density and energy would be high, this process happens very rapidly. Also, as the inverse beta decays progress, the electron density decreases, resulting in a reduction of Fermi pressure allowing the continuance of gravitational contraction. The energy released is carried away by the neutrinos, and in the breakup of iron group elements, the ratio of specific heats could become less than 4/3 which can lead to an implosion along with a large neutrino flux coming from inverse beta decay and electron–positron pair annihilation. In this process, the emitted large flux of neutrinos lifts off the outer layers making the rapid neutron process hastened and adding on a large number of neutrons to the already existing number and makes the entire core '**neutron-rich**'. The dynamics of this collapse is also governed by the opacity of the dense matter in the core to neutrino flux which can lead to rise in temperature and the additional burden of mass and the accompanying processes make the star unstable, making the core implode while its outer shells explode tremendously scattering the debris to the surroundings in a stellar outburst called a '**supernova**' (Irvine, 1978; Harwitt, 1982), yielding a scenario with luminosity at maximum light being of the order of hundred billion times the solar luminosity sometimes getting visible even during the day time.

The Crab supernova which was recorded by the Chinese astronomers was said to be visible for over a month during day and night in 1024 AD. The remnant is visible in all wavelengths from radio to X-rays which indicates the richness of the processes happening in the ejecta and between the ejecta and the interstellar gas clouds (Glendenning N K, 2000).

3.1 Supernovae and Neutron Stars

After the explosion of the star and the moving away of the outer shell in a filamentary structure, will the core set into equilibrium or continue to collapse gravitationally? As mentioned, the core could finally settle down as a highly condensed neutron star. The process of continued collapse takes one back to the discussions of Landau (Landau, 1932) who considered the equilibrium configurations of cold matter and their mass–radius relation. As he pointed out, with the increasing mass, the material is more strongly compressed, and consequently, the Fermi gas which resists the gravitational pull gets squeezed to higher and higher energy per particle, reaching the relativistic level. At this stage, it makes no difference whether the particles are electrons or neutrons, with the linear size of the region being the only parameter of significance. If the configuration contains

Fig. 3.4. Veil nebula-a supernova remnant. The Veil supernova is the remnant of the explosion that could have occurred about 5,000–10,000 years ago. It is located about 1,400 light-years away in the constellation of Cygnus. *Credit*: http://www.astropix.com/html/e_sum_n/veil.

A number of baryons and is of effective radius R, then the Fermi energy of compression per particle is given by $E_F \sim (hc/2\pi R)A^{1/2}$ for neutrons, $E_F \sim (hc/2\pi R)(A/2)^{1/3}$ for electrons. The case of Fermi gas (made of electrons) was discussed in detail by (ChandraSekhar, 1935) for the case of white dwarfs, as already mentioned. He has made a detailed analysis of the equation of state which is valid throughout the entire range of Fermi energies and has further shown that there exists a finite bound as given by the '*Chandrasekhar limit for white dwarfs*' with electron degeneracy pressure counterbalancing the gravitational pull.

On the other hand, if the core mass is over the limit, the core collapse continues to reach the stage when the whole configuration becomes neutron-rich of much smaller size and end up as a more compact stable star, called a '**neutron star**'. Landau's analysis of these configurations were later followed by Oppenheimer, Volkoff, Snyder and Chandrasekhar during 1938–1939. As pointed out by M. Rees (1997),' Walter Baade and Fritz Zwicky of the Mt. Wilson observatory seem to have remarked in 1934 as follows:

"with all reserve we advance the view that a supernova represent the transition of an ordinary star into a neutron star consisting mainly of neutrons"

They further speculated that the supernova explosion was driven by the gravitational energy impulsively released when the star's core collapsed retaining a tiny cinder with closely packed nuclei.

Looking at the evolution of a star of mass of ~6–8 M_{sun} with a low mass core of carbon and oxygen would end up as a white dwarf of mass of about 0.7–1 solar mass, the rest being ejected during the outbursts forming a gaseous envelope of a planetary nebula. Stars which are more massive evolve more rapidly, leading to the formation of a giant along with the production of higher nuclei going up to Iron-56 and few other nuclei, as mentioned earlier. As summarised by Shapiro and Teukolsky (1983) taking into view of the Harrison Wheeler calculations (Harrison *et al.*, 1965), the overall physical picture may be described as follows: 'if the nuclear forces alone determine the structure of nuclear equilibrium, nucleons would accumulate into nuclei of unlimited size. But as there are

Coulomb repulsions which would make such nuclei undergo fission and for low densities, these opposing forces find equilibrium at the value at the atomic number $A = 56$. This would also change if the electrons are relativistic. Then because of inverse beta decay and decreased role of Coulomb repulsion, there will be a greater tendency for the formation of large nuclei. With density reaching $\sim 4 \times 10^{11}$ g/cm^3, the n/p ratio reaches a critical level, and any further increase in density would lead to 'neutron drip' — a two-phase system in which electrons, nuclei and free neutrons coexist to determine the status of low energy. Finally, as the density reaches $\sim 4 \times 10^{12}$ g/cm^3 neutrons provide more pressure than electrons. With neutron gas providing pressure to counter gravity, the equilibrium structure is called a '**neutron star**', as mentioned earlier.

As pointed out by (Glendenning 2000), neutron stars are made up of baryons like nucleons and hyperons, as also can possess cores of quark matter in some cases. During the normal life of a star, the fusion of nuclei from the gravitational compactification producing heat for forming higher nuclei in the periodic table, the end point is reached, with iron nuclei, for exothermic fusion. The core containing heaviest ingredient collapses, releasing enormous amount of energy that leads to supernovae, as was guessed by Baade and Zwicky. Energy released by the solar mass core leads to densities in the core high enough to separate the nuclei into its constituents. The pulse of neutrinos recorded by several large detectors during the observations of supernova 1987 A seems to have confirmed the total integrated energy output over the spherical volume concerned. In these calculations too the Fermi–Dirac statistics played an important role with degenerate neutrons, as it was in the case of white dwarf configuration with degenerate electrons, as was pointed out by Chandrasekhar. As discussed by (Glendenning 2000), burning in the outer shell adds to the core mass which mostly is iron nuclei. Gravity crushes the core rendering electrons to become relativistic and the kinetic energies of these relativistic electrons render a stage for inverse beta decay and if the core mass exceeds the Chandrasekhar limit for white dwarfs, the core induces a rapid implosion in a very short timescale. As the inverse beta decay releases lots of neutrinos, the core gets bloated with a great abundance of neutrino pairs as also due to photodisintegration ($\gamma \rightarrow \nu + \bar{\nu}$), leading to increased Fermi energy. The infalling material rebounds from the stiffened core leading to shock production which travels outwards up to a few hundred kilometers from the centre leading to a decompression wave. At this stage, the core is already full of neutron rich material and the immense gravitational binding energy of the neutron star provides the force for the explosion of a supernova of the material surrounding the central core in a red giant star. Even though the nuclear forces are strong due to the fact that being of short range, neutron stars are bound by gravity which is a long-range force and thus act on all mass energy. As pointed out, the binding energy per nucleon due to gravity is of the order of 160MeV/A, about 10 times greater than the binding energy of nuclear matter which is about 16 MeV/A. Further due to the compressive nature of gravity, the density lies far above the saturation density of nuclear matter. Also, due to the very short-range repulsion of the strong force, the nuclear forces contribute negatively to the binding energy, but it shapes the equation of state which is closely related to the constituents and structure of the star.

Neutron star structure and dynamics is a very engaging topic in astrophysics, particularly in nuclear astrophysics. For a detailed discussion on this topic, one can refer to (Glendenning 2000, Irvine 1977). Basically, this is so because matter at very high densities (like nuclear densities) tend to behave so very differently that one cannot just be satisfied with electrodynamics and hydrodynamics of the Newtonian physics and one needs to bring in additional features. The inclusion of special relativity in the discussion of the degenerate electron gas resulted in a maximum mass, the Chandrasekhar limit for a white dwarf. Could there be similar dynamical limit on the mass of a neutron star with general relativity describing gravity? We shall come back to look at the scenario of the final stages of stellar evolution after a brief discussion of general relativity as the theory of gravitation physics.

Chapter 4

General Relativity the Theory of Gravity

4.1 Introduction

Newtonian theory of gravity describes the gravitational interaction as an action at a distance theory (force travels with infinite velocity), which has been very successful in describing gravity as a force of nature between mass particles. It describes most of the physics we are familiar with locally within the solar system. Also, the equations of Newtonian mechanics are invariant (do not change in the form between different coordinate transformations) under Galilean transformations. Before Maxwell, the phenomena of electricity and magnetism followed different laws as prescribed by Faraday, Oersted, Ampere, and Ohm. Maxwell combined all these laws into a set of four equations that go by his name and gave a theory of electromagnetism in 1865 (Maxwell, 1865) describing well the interaction between charges, currents and magnetic fields. He thus unified the electric and magnetic field interactions and was the first to propose a unified field theory of electromagnetism. Unfortunately, this theory is not invariant under Galilean transformations which are the connecting link between different observers who followed Newtonian mechanics. This indeed could create differences in electromagnetic effects as observed by stationary and moving observers. Further, there were other problems, (1) concerning the role of time and (2) the velocity of light, except to say that it is finite. In 1894, Lorentz (Lorentz 1892) had tried and found the set of transformations which could keep Maxwell's equations invariant for the observers in motion and in 1904 Poincare (1904) had expressed that there should be the same principle of relativity for all physical phenomena including electromagnetism.

Finally, it was Einstein (1905) who integrated all these aspects through the **principle of relativity** by assuming two major conjectures stated as:

1. *light is always propagated in empty space with a definite velocity 'c' which is independent of the state of motion of the emitting body* and
2. *the laws of electrodynamics and optics will be valid for all frames of reference for which the equations of mechanics hold good.*

As the ensuing theory was restricted to observers in uniform motion only, it was called the *special theory of relativity*. The most important aspect of this theory is the fact that the arena for doing physics is the four-dimensional space-time background. The absoluteness of time (Newtonian concept) was removed and in its place, the velocity of light was assumed to be absolute for all inertial observers. The success of this theory mainly came from the invariance of Maxwell's equations for all inertial frames under Lorentz transformations (Lorentz invariance) and the more important derivation of the equivalence of mass and energy ($E = Mc^2$). Incidentally, the theory also explained the negative result of the Michelson-Morley experiment regarding finding the velocity of earth with respect to aether. In 1908, Minkowski (1908) was the first to give a space-time description of the theory using the language of four-vectors which became a standard way for applying special relativity to other areas of physics as well as paved the way for its generalisation. Einstein then worked (1908–1914) on extending the special theory to include gravity and have the laws of physics to be valid for all observers irrespective of their state of motion (the accelerated observers included).

Of all the discoveries of the human mind, Einstein's theory of general relativity is considered to be the most beautiful creation. In fact, it is said that special relativity, which forms a strong basis for modern physics, along with quantum mechanics, was ripe to be discovered at the turn of the nineteenth century, and if not Einstein, Poincare or Lorentz would have developed the theory. On the other hand, the general theory of relativity, which is the epitome of the world of symmetry, assigning freedom from the confines of coordinate systems (observers) to understand the most important of all the fundamental interactions-Gravity, is completely the work of one individual, arising out of thought experiments instead of laboratory experiments or observations, unlike other discoveries in physics.

What are these thought experiments?

4.2 Foundations

In order to generalise the theory of special relativity he assumed what is known as the principle of equivalence, according to which 'for any body, the inertial mass (M_i) (mass that offers resistance to motion) is equivalent to its gravitational mass (M_g) (mass that attracts other masses)'. This in fact is an experimental result, first verified by Eotvos, using a torsion balance in 1889 (Eotvos, 1889), and later confirmed by Dicke *et al.* (1964) to a very high accuracy. From Newton's second law of motion, one has the force acting on a body to change its state, expressed as $F = m_i a$, with m_i the inertial mass and a the acceleration gained by the body. If the force acting on the body is the gravitational force then one will express the same force as $F = m_g g$, m_g being the gravitational mass and g the acceleration due to gravity. From these two one has $a = (m_g/m_i)g$.

Eotvos, performed an experiment with a torsion balance as explained below to check whether this ratio of (m_g/m_i) is different from unity.

Two bobs A and B of different compositions but having the same mass are hung from the ends of a beam suspended at the center by a thin wire. When the beam is in equilibrium, the bodies are subject to two forces:

(i) the gravitational force $m_g g$, acting downwards towards the center of the earth, and
(ii) the centripetal force (inertial force) $m_i f_i$ due to the earth's rotation. Equating the components of the forces acting in the directions $\sim i$, and $\sim k$ (See Fig. 4.1) one gets, along the vertical,

$$\vec{i} = l_A\left(-m_i^A f_i + m_g^A g\right) = l_B\left(-m_i^B f_i + m_g^B g\right). \tag{4.1}$$

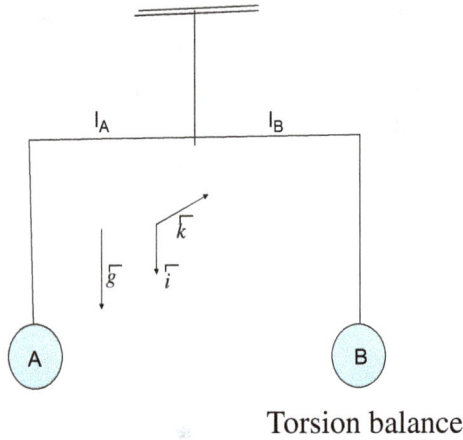

Torsion balance

Fig. 4.1.

(iii) Along the horizontal direction $\sim k$, any unbalanced force should induce a torque T, as given by

$$T = (l_A m_i^A f_k - l_B m_i^B f_k). \tag{4.2}$$

Using this in (2.1) above, one gets for T the expression

$$T = l_A m_i^A \left[1 - \frac{m_g^A m_i^B}{m_g^B m_i^A} \right] = l_A m_g^A f_k \left[\frac{m_i^A}{m_g^A} - \frac{m_i^B}{m_g^B} \right].$$

As no torque was found in repeated attempts $T = 0$ implies $m_i^A / m_g^A = m_i^B / m_g^B$, showing the equivalence of the two masses which when used in the force equations yield $m_i^A a = m_g^A g$ or $a = g$.

This indicates that the acceleration field is the same as the gravitational field. Thus, Einstein came to the conclusion that accelerated observers are equivalent to observers in a gravitational field. Equivalently a gravitational field can be replaced by an acceleration field. Then how can one visualise gravity in accelerated frames? Using the freely falling elevator example Pirani (1957) illustrated the discussion as follows.

Consider that the person in the elevator drops two coins side by side as shown in (Fig. 4.2(left)). Neither of them falls to the bottom of the lift

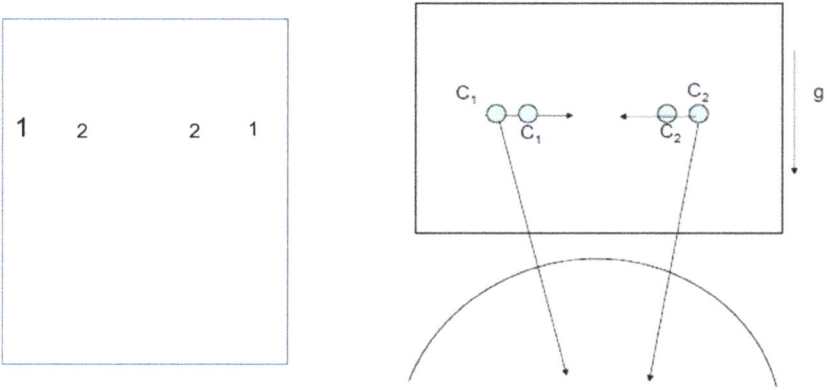

Fig. 4.2.

but as the elevator descends, the two coins appear to come closer to each other.

Why is that so? Looking at the diagram on the right, one can see that both coins are following their trajectories (world lines) which indeed meet at the center of gravity of the earth and thus converge towards each other.

This would clearly show that there is a force acting between the two coins, apart from gravity. This is referred to as the *relative acceleration*, which cannot be taken away. In order to understand this mathematically let us consider two particles falling freely in a gravitational field φ as shown in the diagram (Fig. 4.2).

Let us denote their trajectories by λ_1 and λ_2 and a connecting vector η^i between the two. The equations of motion for the two particles, with $\partial_i = \frac{\partial}{\partial x^i}$ are

$$\ddot{x}^i = -(\partial^i \varphi)_P = -\delta^{ij}(\partial_j \varphi)_P \tag{4.3}$$

$$\ddot{x}^i + \ddot{\eta}^i = -(\partial^i \varphi)_Q = -\delta^{ij}(\partial_j \varphi)_Q \tag{4.4}$$

As η^i is very small, one can use Taylor expansion in the second equation to get

$$(\partial_j \varphi)_Q = (\partial_j [\varphi (x^i + \eta^i)] = (\partial_j [\varphi(x^i) + \eta^k(\partial_k \varphi)$$
$$= (\partial_j \varphi)_P + \eta^k \partial_k \partial_j (\varphi)] \tag{4.5}$$

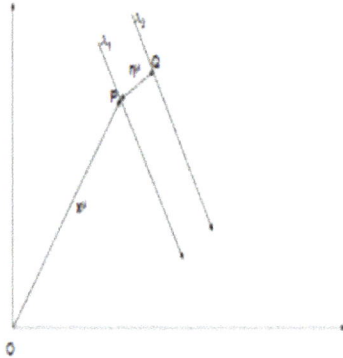

Using these expressions in the equations of motion and simplifying one gets the relation,

$$\ddot{\eta}^i = -\eta^k \partial_k \partial^i (\varphi) \tag{4.6}$$

the equation for relative acceleration between the particles in the Newtonian framework. In fact, the 'tides' that happen in oceans are due to this relative acceleration between the moon and earth both falling freely in the gravitational field of the sun. We shall consider this again later to obtain the corresponding equation in general relativity through the study of geodesic deviation in a curved spacetime.

4.3 Light and Gravity

As Einstein had already expressed the equivalence of mass and energy ($E = mc^2$), and as light is a form of energy, what would be the effect of a change in the gravitational field on a pulse of light?

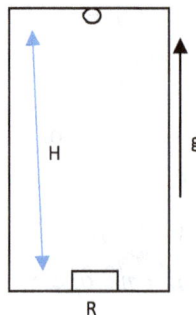

Consider a freely falling space ship of height H as shown which can be equivalently considered as being uniformly accelerated with acceleration (+g). Consider a light-emitting device (O) at the roof of the ship and a receiver (R) at the bottom. If O emits a pulse of light it will reach R in H/c secs. During this time the ship would have acquired a velocity (gH/c) m/s in the upward direction. If the emitted frequency is λ_e and the observed frequency is λ_o there will be a Doppler shift such that

$$\frac{\lambda_o - \lambda_e}{\lambda_e} = -\frac{v}{c} = -\frac{gH}{c^2}. \tag{4.7}$$

In other words, the frequencies of the two pulses are related as given by

$$\nu_e = \nu_o \left(1 - \frac{gH}{c^2} \right). \tag{4.8}$$

This implies that the observed frequency is larger than the emitted frequency and their relative values depend upon whether the receiver is moving towards the source or away from it. As $(-gH)$ is the difference in the gravitational potential $(\delta\varphi)$ between the top and the bottom of the ship, one can conclude that the frequency of light increases or decreases depending upon whether the light is travelling from a higher potential to a lower potential or vice versa. Using this argument Einstein predicted that light from the sun received on earth should be red shifted and this shift in spectral line is called the *gravitational red shift* or Einstein's red shift defined by the relation

$$\nu_o = \nu_e \left(1 + \frac{\delta\varphi}{c^2} \right), \quad \delta\varphi = \varphi_o - \varphi_e \tag{4.9}$$

giving the frequency shift,

$$\frac{\delta\nu}{\nu} = \left(\frac{\nu_o - \nu_e}{\nu} \right) \approx -\frac{GM}{c^2} \left(\frac{1}{R} - \frac{1}{r} \right) \tag{4.10}$$

R being the Solar radius and r the radius of the earth's orbit. This red shift was rediscovered by Mossbauer in 1958 in γ-ray spectroscopy (1958) and is known as the Mossbauer effect. In 1960, Pound and Rebka (1960) experimentally verified (Pound and Rebka 1962) the result with sufficient accuracy using gamma rays. This effect of gravity on light had led Einstein to predict the bending of light rays in a gravitational field by assuming that the frequency shift in differing potentials is due to the fact that the clocks at two different locations go slower by the factor $(1 + \delta\varphi/c^2)$, with relative gravitational potential difference $\delta\varphi$. This in turn, changes the velocity of light locally, by the same factor. He therefore concluded that if c_0 represents the velocity of light at the origin of coordinates then the velocity at a place with gravitational potential $\delta\varphi$ would be equal to $c = c_0(1 + \delta\varphi/c^2)$. Consequently, using Huygen's principle he derived in 1912 that a light ray coming from infinity and passing through the gravitational field of a point source would suffer a deflection towards the source by the factor $(2GM/Rc^2)$, M being the mass of the source and R the distance of closest approach by the light ray and c the velocity of light in vacuum. It is in fact lucky that the experimental verification of this result planned by Finley Freundlich could not be done going to Russia due to the ensuing first world war and thus could not be verified. However, later in 1916 with the application of the full theory of general relativity, the correct value was found to be $(4GM/Rc^2)$ twice the number expected in 1912 and was verified by the team of Eddington in 1919, which was considered as the best experimental verification of the new theory of gravity.

The most important feature that was apparent at this time (1912) to Einstein was that the extension of special relativity to include gravitation could not be done on the Minkowski flat manifold due to the following space-time diagrams. If an observer O_1 sends a beam of light (signal of frequency ν_1) to a colleague O_2 who receives it after a time lapse Δt_1, in Minkowski space this would be represented as in Fig. 4.3.(left):

If the signal is of N cycles then the duration of emission $\Delta t_1 = 2\pi N/\nu_1$. The received signal is of frequency ν_2 which is greater than ν_1. Consequently, this would mean $\Delta t_2 = 2\pi N/\nu_2$ should be greater than Δt_1 which needs to be represented in a diagram as given on the right-hand side of Fig. 4.3. Thus, it can be concluded that the space-time diagram cannot be on a Minkowskian background but has to be on a curved manifold, where the photon trajectories can be depicted as light curves just as in case of

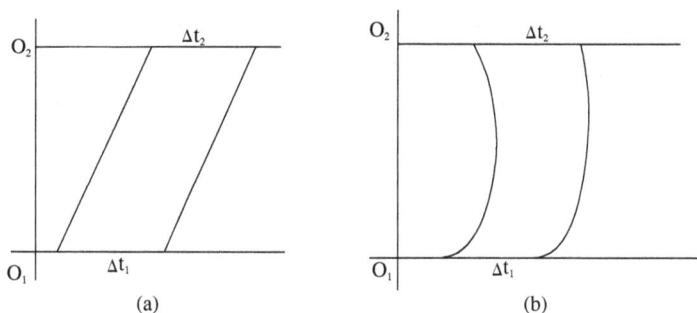

Fig. 4.3.

accelerated test particles. Apparently, he had realised this and later expressed as follows: *If all accelerated systems are equivalent then Euclidean geometry cannot hold in all of them and thus I realized that Gauss's theory of curved surfaces might hold the answer and that Gauss surface coordinates must be profoundly significant.* This had made him look for an alternative geometry to use as the background structure for the new theory. Luckily his mathematician friend Marcel Grossmann came to rescue and introduced Einstein to Riemannian geometry (a non-Euclidean geometry) of positive curvature and its further developments by Ricci, LeviCivita and Bianchi which successfully provided the canvas for general relativity in 1914. Why Riemannian geometry?

This is a generalisation of Gauss's curved geometry of a 3-surface to n dimensions. It is well known that if one draws a triangle on the surface of a sphere with one of the poles and two points on the equator, the sum of the three angles of such a triangle is greater than π which is non-Euclidean. Further if one wants to draw an Euclidean triangle on the surface of a sphere it can be only an infinitesimally small triangle just around a given point, where one can have an extremely small flat region. This gave the possibility for Einstein to introduce the postulate *the principle of equivalence* which in its strong form, says 'at every point on a curved surface one can introduce in its immediate infinitesimal neighbourhood flat space coordinates (Lorentzian geometry) where special relativity holds good'. More technically the principle says that 'one can set up a local Lorentz frame in every infinitesimal neighbourhood of any point on the given surface where special relativity holds good. As we are dealing with the four-dimensional space-time geometry, the surface referred to is a three surface.

While wanting to generalise the special theory of relativity to include accelerated observers, Einstein wanted to have *all laws of physics to be covariant in all frames of reference*. For this, he introduced the second postulate which is known as the *principle of general covariance*. As Grossmann pointed out to Einstein, Riemannian geometry is built on what are called tensors, which by definition are always covariant under all coordinate transformations.

A tensor of rank zero is a scalar and of rank one is a vector. Thus, by adopting the Riemannian geometry as the base manifold for general relativity, Einstein freed the theory from the confines of coordinates which implied the freedom from observers in any state of motion. One could express this figuratively as: The principle of equivalence generalises the special theory 'by fact' where as the principle of covariance takes it 'by form'.

Now that one is on a curved surface the trajectories of particles as well as of photons (light) are no longer straight lines but the geodesics of the given geometry which indeed are the shortest path between any two points. (extremals of the given metric) $ds^2 = g_{ij} \, dx^i \, dx^j$. For a detailed description of the theory with mathematical details, one can refer to any of the following books (Weinberg, 1972; Misner *et al.*, 1973; Rayd'Inverno, 1992; Prasanna, 2017).

4.4 Applications

What are the initial successes of the theory? A theory is considered to be successful if it satisfies the following criteria:

(i)   Theory should explain some observation which could not have been fully explained by the earlier theory and

(ii) It should predict some new effect which has to be proven true.

One of the problems in celestial mechanics was the precession of the perihelion of planetary orbits, particularly of Mercury (which is closest to the sun) which had a discrepancy between the observed value and the theoretically predicted value on the basis of Newtonian theory, (taking into account the gravitational effects from the other planets) the calculated value of 5557 ± 0.20" is less than the observed value of 5600 ± 0.41", by about 43" arc per century.

According to general relativity wherein the celestial orbits are given by the particle trajectories-geodesics of the underlying space-time, the orbit of Mercury around the Sun is given by the modified Newtonian equation

$$\frac{d^2u}{d\varphi^2} + u = \frac{m}{h^2} + 3mu^2, \quad u = 1/r, \tag{4.11}$$

φ being the azimuthal angle and h the angular momentum. The equation can be solved perturbatively as it is an equation for a conic section with relativistic correction coming from general relativity. Using the expansion

$$u = u_0 + \varepsilon u_1 + \cdots,$$

one can find the solution to be

$$u = u_0 + \varepsilon \frac{m}{h^2}\left[1 + e\varphi\sin\varphi + \frac{e^2}{6}(3 - \cos 2\varphi)\right], \tag{4.12}$$

where $u_0 = \frac{m}{h^2}\left[1 + e\cos\varphi\right]$, is the classical Newtonian solution with e the eccentricity of the orbit. Of the various relativistic correction terms, only the second term on the right-hand side is of significance as that is the only term increasing with every revolution of 2π, and thus the final solution for the orbit may be written as

$$u = (m/h^2)[1 + e(\cos\varphi + \varepsilon\varphi \sin\varphi)]. \tag{4.13}$$

In view of the fact that the term $\varepsilon\varphi$ is very small ($\varepsilon\varphi \ll 1 \rightarrow \sin \varepsilon\varphi \cong \varepsilon\varphi$) the final solution can be taken as

$$u = \frac{m}{h^2}[1 + e\cos(1-\varepsilon)\varphi]$$ (4.14)

which describes an ellipse with a secular period $2\frac{\pi}{1-\varepsilon} \cong 2\pi(1+\varepsilon)$. This indicates that the orbit is not a closed ellipse but one with a slightly changing location of the perihelion in every orbit. Thus, the correction is $2\pi\varepsilon = 24\pi^3 a^2/c^2 T^2(1 - e^2)$, where a is the semi-major axis, T the period and e the eccentricity of the orbit. Using the known values for these orbital parameters one can find that the correction amounts to ~42.9" of arc as was required.

Thus, Einstein's theory satisfied the requirement of explaining a previously unexplained lacuna in the Newtonian theory of orbital mechanics. Regarding a new prediction by the theory, one can look at the photon trajectories as it then completes the study of trajectories of all particles including the one with zero rest mass (photon).

4.5 Photon Trajectory and Bending of Light Rays

Even before the formulation of general relativity, Einstein had predicted the effect of gravity on light as defined by the gravitational red shift saying that a pulse of light going from a higher gravitational field to a lower one would have a frequency shift. This also was one of his gedanken experiments as mentioned earlier. The important fact that was realised was that the photons move along the null geodesics of the underlying space-time. After Einstein formulated the theory in 1914 and obtained the corresponding field equations for gravity, Schwarzschild was the first to obtain an exact solution for the field of a spherical mass distribution. The exterior field is given by the well-known Schwarzschild solution:

$$ds^2 = \left(1 - \frac{2m}{r}\right)dt^2 - \left(1 - \frac{2m}{r}\right)^{-1} dr^2 - r^2 d\theta^2 - r^2\sin^2\theta d\varphi^2.$$ (4.15)

Considering the Lagrangian corresponding to (4.15) one can write in terms of proper time derivatives of the set of equations:

$$c^2(1 - 2m/r)\dot{t} = E, \quad r^2\dot{\varphi} = h$$ (4.16)

with E denoting the energy and h the angular momentum of the particle.

For the case of null geodesic $ds^2 = 0$, gives the third integral

$$(1 - 2m/r)^{-1} \left[E^2/c^2 - \dot{r}^2 \right] - (h^2/r^2) = 1. \tag{4.17}$$

Rewriting in terms of the parameter $u = 1/r$, and the angular coordinate φ, the equation takes the form

$$(du/d\varphi)^2 = 2\, mu^3 - u^2 + \text{constant}. \tag{4.18}$$

Differentiating this with respect to φ one gets the equation

$$d^2u/d\varphi^2 + u = 3\, mu^2. \tag{4.19}$$

which in the absence of any gravitating source ($m = 0$) yields,

$$d^2u_0/d\varphi^2 + u_0 = 0,$$

an equation to a straight line through the solution $u_0 = \frac{1}{d}\sin(\varphi - \varphi_0)$, d being the closest approach of the ray to the origin of coordinates as φ completes a revolution from 0 to π. Solving (4.19) perturbatively, assuming $u = u_0 + 3m\, u_1$, one can get the equation for u_1

$$d^2u_1/d\varphi^2 + u_1 = u_0^2 = \sin^2\varphi/d^2,$$

whose solution is given by

$$u_1 = (1 + C \cos \varphi + \cos^2 \varphi)/3d^2, \tag{4.20}$$

and the final solution therefore is given by

$$u = \sin \varphi/d + (m/d^2)\,(1 + C \cos \varphi + \cos^2 \varphi). \tag{4.21}$$

To analyse the path of a light ray coming from a distant star and passing the limb of the sun, one can consider the asymptotes to the ray at the source and at the observer and taking the limits for $r \to \infty$, $u = 0$, for the angle $\varphi = -\varepsilon_1$, and $\varphi = \pi + \varepsilon_2$ and get the equations

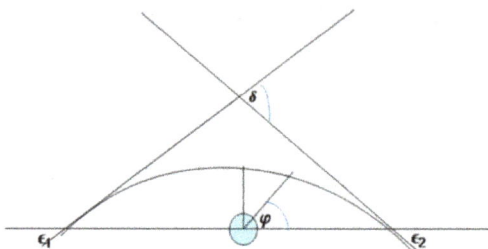

Fig. 4.4. Bending of light ray at the solar limb.

$$-\varepsilon_1/d + m(2 + C)/d^2 = 0, \quad -\varepsilon_2/d + m(2 + C)/d^2 = 0,$$

which gives $\varepsilon_1 + \varepsilon_2 = 4m/d$.

The accompanying Fig. 4.4 clearly shows that the deflection angle $\delta = \varepsilon_1 + \varepsilon_2$. For the light ray under consideration, grazing the sun's limb, the deflection angle turns out to be $\delta = 4m/R$, R being the outer solar radius. Using the known values for the mass and radius of the sun, one finds the deflection angle to be $\cong 1.75$ *arcsecs*. This prediction of the theory was tested by Eddington and his team in a total solar eclipse in 1919 and found to be exactly right. After the advent of radio astronomy, one need not have to wait for total solar eclipse as the radio waves from a distant Quasar also go through the same phenomena of bending by some intervening heavy star. However, one has to know the interstellar electron column density in the line of sight as there could always be the dispersion of radio waves through electron–photon interactions. Thus, it was found that the theory of general relativity successfully satisfied both criteria for any new theory.

The most important consequence of light bending is the discovery of the phenomenon of *Gravitational lensing*, first predicted by Einstein himself and later discussed by several other astrophysicists. We shall consider this later after going through the consequence of the effects of the general theory of relativity in the scenario of gravitational collapse as introduced by Hoyle and Fowler in the early sixties in the context of the energetics of Quasars and Active galactic nuclei. Before moving on to consider the physical aspects, it is relevant to consider how the physical quantities are related to the geometrical quantities as depicted in the relativistic situation. As the theory of general relativity is generally covariant one can use any system of coordinates while dealing with its application to physical situations particularly dealing with electrodynamics or hydrodynamics. There are two particular frames of interest which can be useful in future applications, namely (1) local Lorentz frame and (2) locally non-rotating frame.

4.6 Locally Inertial Frames

The principle of equivalence which is one of the basic features of GR, assures the existence of locally inertial frames as according to the principle at any point in a differential manifold one can have special relativity in its close neighborhood and thus an inertial frame known as local Lorentz frame (LLF). As the field equations and the associated fluid equations are fully covariant (tensorial formulation) all the physical parameters are in geometrised components which need to be expressed in local Lorentz frames as the components in that frame are the physically observable representations. In every situation, one can introduce a locally inertial frame which helps in converting the geometrical parameters in terms of physical parameters. For any given geometry one can relate the geometrical quantities to physical quantities through the introduction of the local Lorentz frame λ_a^i where the alphabets i, j, k etc denote the vector indices and the letters a, b, c etc. stand for the coordinate indices. The λs are defined through the normalisation condition $\eta_{ab} = g_{ij} \lambda_a^i \lambda_b^j$, g_{ij} and η_{ab} being the metric tensor and the Minkowski tensor (flat space) respectively. With this, all the geometrical quantities are related to the corresponding physical quantity as given by $A_i = \lambda_i^a A_a$ and $A_{ij} = \lambda_i^a \lambda_j^b A_{ab}$. As both the metric tensor and the Minkowski tensor are invertible the inverse matrix λ_a^i satisfies the relation $\lambda_a^i \lambda_j^a = \delta_j^i$. It is important to note that in choosing the local Lorentz frame one has six degrees of freedom as the equations determining the sixteen non-zero components of them are related by only ten equations. The six degrees of freedom are associated with the rotation group $4C_2 = 6$ rotations — the three Eulerian rotations (space-space) and three boosts (space-time rotations). These locally inertial frames are very useful while doing the hydro dynamics and electrodynamics in GR formalism as these connect the physically meaningful point functions (pressure, velocity etc.) that are defined locally with their global (geometrical) definitions valid over all underlying space-time.

4.7 Locally non-Rotating Frames (LNRF)

Rotation is a very important concept in dynamics and particularly so in the context of general relativity where the most important general spacetime is given by the axisymmetric stationary space time (ex Kerr space time). Also known as zero angular momentum observers (ZAMO), these coordinates were introduced by Bardeen (1970), (Bardeen *et al.*, 1972) refer to an observer who moves with the angular velocity $\omega = -g_{\varphi t}/2g_{\varphi\varphi}$. Misner *et al.* (1973). The significance of such coordinates (observers) comes in the context of the discussion of the inertial frame dragging as such observers do not see the rotation associated with the spacetime as they move along with the same angular velocity. The central source is thus assumed to be static in the commoving frame, although in uniform rotation. This uniform rotation is the configuration that minimises the total mass energy at a specified baryon number and angular momentum (Hartle and Sharp, 1967). Later during the discussion of the non-gyration of particles in the Kerr geometry, we shall use these coordinates to emphasize the concept of inertial frame dragging which does not seem to allow the gyration of particles within the ergosphere in BL coordinates.

Chapter 5

Radio Universe and its Role in Astrophysics

5.1 The Beginnings

In the 1930's as electronic communication facilities increased, the Bell telephone company, a leader in that technology noticed a problem. They found that there was a continuous static interference with short-wave transatlantic voice signals and assigned the problem to Karl Jansky a radio engineer. Jansky, using a specially constructed antenna made meticulous observations and noticed that the particular disturbing signal appeared in a periodic manner-once every sidereal day as pointed out by his friend astrophysicist A.M. Skellett. Jansky then compared his recordings with optical astronomical maps and came out with the *suggestion that the disturbing radio signal was not from any manmade source but must be coming from the constellation Sagittarius-the center of the Milky Way galaxy.* Jansky announced his discovery in 1933 and *this was the birth of Radio astronomy.* However, no progress occurred till 1937, when Grote Reber built a 9 m parabolic antenna and conducted a survey of cosmic objects emitting at radio frequencies (as described later by Kellerman (1999)). Unfortunately due to the ongoing second world war in the late thirties, developments in radio astronomy had to wait till the late forties. It is interesting to learn that even the war deeds did help in a new discovery, (when it was wrongly presumed that the jamming of British radars was by the enemy, whereas detailed investigations revealed the jamming was due to the appearance of large sunspots). Similar to Jansky's discovery of radio emission from the milky way, the discovery of radio emission from the Sun was realised in 1942. As the process of synchrotron emission due to the electrons circling around the magnetic field lines was known, the solar radio emission was thus identified with the plasma processes and the variations associated came from the plasma oscillations. This lead to the developments in solar radio astronomy which is a rich source of information about the different aspects of solar emissions from the quiet sun to solar radio bursts. Apart from the galactic centre, another important source of radio signal observed in the fifties was that from the source Taurus A, optically identified as the Crab nebula which basically is the remnant of the supernova 1054 AD, and its expanding shell. Apart from this in 1954 W. Baade and R. Minkowski discovered the second strongest radio source, which could have been detected even if it had happened to be at far greater distance as radio telescopes are extremely sensitive to weak signals.

As M. Rees (1997) points out, at that time it was felt that the strongest source of cosmic radio signals came from the galactic center and the Crab nebula, whereas the second strongest source observed in the early fifties was that from a remote galaxy, discovered by W. Baade and R. Minkowski. As it was realised that the number of radio sources appeared to be far higher than what was discovered optically, the interest moved mainly towards the cosmological implications of the new findings.

Even though optical astronomers knew about the emission line nuclei in galaxies in the early twentieth century, real advances in this area had to wait till the nineteen fifties. The bright and dark lines of NGC 1068 were confirmed by Slipher (1917) with spectra taken in 1913 at Lowell Observatory. In 1917, he obtained a spectrum with a narrow spectrograph slit, and found that the emission was spread over a substantial range of wavelengths.

A systematic study of galaxies with emission lines from their nuclei came only in 1943 with Seyfert getting the spectrograms of six galaxies with nearly stellar nuclei showing emission lines superimposed on a normal G-type (solar-type) spectrum. Optical identifications of discrete sources were finally achieved by Bolton, *et al.* (1949). They further identified Taurus A with the Crab Nebula supernova remnant and Virgo A with

M 87, a large elliptical galaxy with an optical jet and Centaurus A with NGC 5128, an elliptical galaxy with a prominent dust lane. Thus, started a working partnership of optical and radio astronomers (Greenstein & Schmidt, 1964).

The 1950s saw progress in radio surveys, position determinations, and optical identifications. In order to increase the sensitivity and the resolving power, Martin Ryle and his team (Smith and Elsmore, 1950) developed the interferometric studies, and a class of sources distributed fairly uniformly over the sky was shown in the survey made by this team. These studies, (Ryle's survey) in fact stood as a basis for the analysis of the cosmological models. Smith (1950) obtained accurate positions of four discrete sources, Tau A, Vir A, Cyg A, and Cas A., which enabled Baade and Minkowski (1954) to make optical identifications of Cas A and Cyg A in 1951 and 1952. This helped in working out the redshifts which gave a velocity of 16,830 km/sec for Cyg A implying a large distance of about 31Mpc, for the assumed Hubble constant, $H_0 = 540$ km/sec/Mpc. *The large distance of Cyg A implied a large luminosity* ~8×10^{42} *erg/sec in the radio, much larger than the optical luminosity of* 6×10^{42} *erg/sec* and enormous compared to any of the earlier known sources. Observing such high energy emissions from point sources in the cosmos was a great delight to the astrophysicists and to understand the energetics of such sources was a big challenge. For a historical introduction to these studies, one can refer to the article 'Cavendish astrophysics', at https://www.astro.phys.cam.ac.uk/.

5.2 Quasars

Quasi-stellar radio sources or Quasars are the optical objects of stellar size associated with extremely large radio emissions. In 1960, Mathews *et al.* were the first to announce the discovery of the object 3C 48, and subsequently in the next two years, they announced the discovery of three more objects of similar nature, 3C 286, 3C 147 and 3C 196. However, the sources were enigmatic as no understood emission mechanism could be associated with them. Luckily, Schmidt in 1963 identified a similar radio object associated with a very bright stellar companion whose spectrum could be identified with a large red-shift. Just after this, Mathews along with Greenstein could explain the spectrum of 3C 48 with even a larger redshift. These results confirmed the interpretation that the redshifts of these sources as gravitational effects of either very dense or very massive objects from far away super luminous objects in galaxies. Using the method of lunar occultation (Hazard *et al.*, 1963) had shown that the source 273 is a double source with a separation of the order ~19.5", measured to an accuracy of around 1". Remarkable observation of the source 273 by A Sandage had shown that the components coincide almost precisely with a thirteenth mag. star towards the end of a faint jet. It was indeed shown that this source was one of the several radio sources where one of the components coincides with an optical star. Investigations relating to the proper motion of sources when applied systematically to 3C 273 by the Yale team under H.Smith (Jeffrys 1963) reportedly indicated that no proper motion could be detected and that the source must be located beyond 20KPc placing it beyond the Milky way a finding also confirmed by Lyuten (1963). If one assumes the redshifts of these sources are cosmological, and the Hubble constant to be ~100km/sec/Mpc their distances work out to be ~1,100 Mpc (3C 48) and 474 Mpc (3C273), with their absolute visual magnitudes turning out to be about –25 and –26 respectively making them the brightest objects known till that era (Greenstein & Schmidt, 1964). The source 3C 48 has a stellar image except for faint reddish wisps with the star being about 3' north of the center of these features, with irregular intensity distribution as reported by Mathews and Sandage (1963). Also as was reported by Munch and Greenstein and Munch (1961) there exists a nearly featureless continuum extending far into ultraviolet and of weak, broad emission lines which are not found in normal or peculiar stars. During the early sixties (1960–1964) several observations of these and similar radio objects were carried out and as reported by Sandage (1964) identification with optical objects and the intensity variations of some of the main ones can be considered a significant development in the study of radio astronomy. The data accrued by then showed that the optical flux had changed by a factor of 1.4 at all the observed wavelengths over a period of ~600 days, which amounted to a change in power output of ~4×10^{44} ergs/sec. Among the various plausible reasons for these variations it was suggested that some could be as follows:

1. If the optical flux is due to synchrotron emission, then a high-frequency cut-off of relativistic electrons by new injection leads to energy decay,
2. Variation of the electron density of a high temperature ($T_e \sim 50K$) gas radiating by free-free and free-bound emission or
3. Storage and energy release as in pulsating variables.

With such enormous energy emissions, it obviously became a challenge for the astrophysicists to develop models to describe the origin and sources for the energetics of these objects. As pointed out by Hoyle and Fowler (1963) and further worked out by (Hoyle *et al.*, 1964) Gravitation could be the main reason, though the observations do not prove the energy per unit mass exceeds one part in ~112 obtainable by thermonuclear combustion.

The process now well known as the gravitational collapse was studied in detail by Landau, Oppenheimer *et al.* and Chandrasekhar in the1930s where general relativity played an important role. This part of the discussion will be taken up later while discussing the scenario of continued gravitational collapse. One of the most important techniques devised to study the extrinsic structure (angular size) of Quasars was the method of lunar occultation. During the transit of the sources, it is obvious that the source could disappear behind the moon and this feature has been noted in detail by several observers. Hazard *et al.* (1963) made a detailed study of the source 3C 273 using this technique and confirmed its double nature with a separation of the order ~19.5′ as mentioned earlier. Determining the position of each of the components to an accuracy of ~1′ it was noted that the component B has a diameter of ~3′. One of the early images of the quasar was taken by A Sandage (1963/1964) with the Palomar 200" telescope, while Schmidt (1963) had taken its spectra with the prime focus spectrograph at dispersions of 190 and 400 $\mathring{A}/mm$, showing broad emission lines resembling the Balmer series having unexpectedly large red shift (for that era) of $z = 0.158$. Figure 5.1 shows the photograph of the quasar taken by ESA &NASA using the wide field camera 2 of the Hubble space telescope in 2013.

The other quasar of great interest and which also happens to be the first one identified with an optical counterpart is the radio source 3C 48,

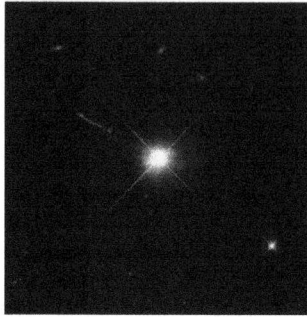

Fig. 5.1. The Quasar resides in a giant elliptical galaxy in the constellation Virgo and estimates show that its light has taken almost 2.5 billion years to reach the earth though it is the closest quasar. The continuum is blue with a nearly flat energy distribution with its apparent brightness being $m_v = +12.6$ yielding a flux of the order $\sim 3.5 \times 10^{-25}$ (erg/cm^{2}sec) (c/s)$^{-1}$ (Oke, 1963).

first identified by Mathews *et al.* (1960). As discussed by Greenstein and Schmidt (1964) the observed redshifts of these and a couple of other quasars of matching red shifts clearly indicate that of the three possible mechanisms for red shift of distant sources, Doppler effect from high-velocity stars and gravitational redshift are both unlikely sources except to confirm the extra galactic nature of them. The third possibility cosmological redshift which occurs due to the expansion of the universe is the most likely explanation and as their emission spectra reveal that the gaseous nebula, the source for the line emission, could be about 1 pc radius for 3C273 and ~10 pc larger for 3C 48. It has also been noticed that whereas the non-thermal radio spectrum is indicative of the emission coming from outside the gas cloud, the light variations require a smaller source for the optical continuum, especially for 3C 48. However, these ideas do seem to face some difficulties concerning their ages. More discussions on these aspects may follow later in the context of cosmological models of the universe.

Concerning some other observational features two important results which require some attention are the process of gravitational lens to be

(1) bending of radio waves from stellar occultation due to the gravity of the star and

(2) observation of binary and multiple optical images of quasars due to the relativistic light bending.

These aspects will be discussed later in the context of the gravitational lens effect.

5.3 Pulsars

One of the interesting things one notices while looking at the night sky is the twinkling of stars. What is it due to? Light waves being electromagnetic waves interact with the plasma of the solar wind that covers the interplanetary space. When electromagnetic waves pass through a plasma, the waves interact with the ions and electrons of the plasma and suffer intensity variation due to wave–plasma interaction. In fact, the passage of the waves depends upon the plasma frequency $\omega = \sqrt{4\pi n_e/m}\, e$ which is a function of the electron density n_e, and the electron charge e and its effective mass m. As the interplanetary medium contains tenuous plasma of the solar wind with varied electron density both across space and time, the passing waves suffer intensity variation at the observer on earth and thus stars appear twinkling. The same is true for radio waves too and the phenomenon is called *scintillation*.

The scintillation pattern varies between point sources and the extended sources, with the intensity fluctuation for point sources being deeper. Scintillation in radio frequencies due to the ionosphere was first observed by Hewish (1951) and subsequently, in 1954, he reported observing the same effect from the radio source in Taurus. The interplanetary scintillation programme (IPS) became an important study for radio astronomers in the 1960s.

Hewish built the Inter Planetary Scintillation (IPS) array at the Mullard Radio Astronomy Observatory in Cambridge, consisting of 2,048 dipoles over almost five acres of land. This array was used to constantly survey the sky at a time resolution of about 0.1 secs, which was in fact considered the best among all the then-operating radio telescopes.

Jocelyn Bell joined this group (Cambridge IPS group) as a research student in 1966 to work on the scintillation array mentioned above and worked meticulously on the project. As there were no computer-aided

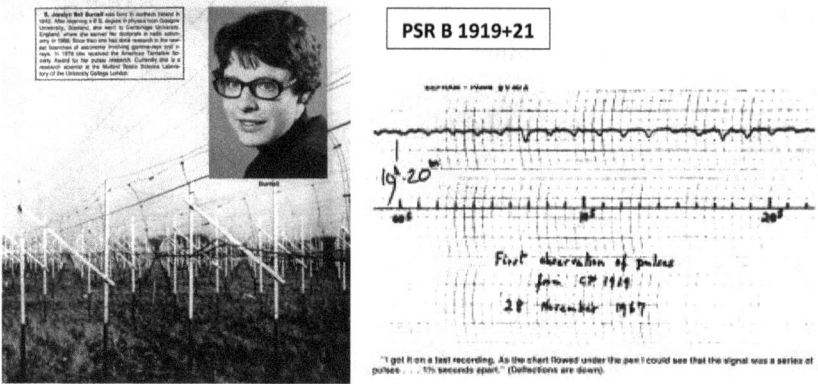

Fig. 5.2.

recording systems she had to run through pages and pages of paper recordings of the arriving signals to look for scintillation patterns. In 1967 November, she noticed a 'bit of scruff' as she called it, on the recorder showing a pulse repeating every 1.33 sec. Checking back the records to a couple of months back she noticed exactly a similar pattern of pulses with regular periodicity.

When she brought this feature to the notice of her supervisor A. Hewish, he apparently did not believe that it could be celestial in origin. Ms Bell however was so convinced, she started checking the old recordings in different parts of the sky and after a long, tiring search of almost a couple of km length of the chart paper she found three more objects of similar pattern but with different periodicity. All of them showed extremely regular periodicity of pulses and thus came the discovery of **Pulsars**. This work was published in the journal 'Nature' in Feb. 1968, Hewish *et al.* (1968). Within a year of this announcement, several more Pulsars were discovered almost all within the realms of our galaxy with periods varying from about 1.05 secs (slower one) to 0.033 secs (33 millisecs, the fastest for that era). The discovery of such regulated cosmic clocks immediately attracted the theorists to model the object using the then-known and established facts of stellar evolution. It was clear that the only way a regular period can be maintained is through the rotation of the emitter. The only confirmed compact objects known were the white dwarfs. A large amount of work was carried out to calculate the oscillation periods of white dwarfs and the idealized theory led to the period greater than about 2 secs which was not satisfactory. Further, the discovery of the Crab pulsar with 0.033 secs period clearly ruled out the white dwarf scenario. One had to look further for more compact objects which would stay dynamically stable under extremely fast rotation. The answer turned out to be the neutron star for which the equatorial velocity of rotation at P = 0.033 sec is small enough and the gravitational acceleration is great enough to counter the centrifugal force (Zeldovich and Novikov 1972).

Thomas Gold of Cornell university was the first to suggest that Pulsar is a rotating neutron star (Gold 1978). The existence of neutron stars had already been proposed as the remnant of a supernova by Baade and Zwicky (1934).

Woltjer (1964) had suggested that the collapsing core would have a very large magnetic field due to the conservation of magnetic flux and a main sequence star could end up with surface fields of the order 10^{14} to 10^{16} G, when it reaches the neutron star stage. In 1967, F. Pacini (1968) had suggested that such a rotating neutron star with magnetic field could emit radiation. Gold's suggestion (Gold, 1978) was that the strong magnetic fields and high rotation speeds allow relativistic velocities to be set up in any plasma in the surrounding magnetosphere, leading to radiation in the pattern of a rotating beacon. Ginzburg and Zhelezniakov (1975) suggest in their review of pulsar emission mechanisms that a pulsar is a rotating neutron star with a strong dipole magnetic field with the rotation axis and magnetic axis being misaligned and having a partially corotating plasma magnetosphere. This could have a localised radiation source, perhaps above the magnetic pole and if the radiation is beamed, successive radiation pulses spaced by star's period of rotation would be observed on earth. This rotating neutron star model also clarifies the earlier speculation of Baade and Zwicky regarding the association of cosmic rays with supernovae, as seen in the case of the Crab pulsar.

As Rees puts it, the gradual slowing down of the Crab pulsar could be the evidence for the star's spin being responsible for the radiation and into a wind of particles that keep the Crab nebula shining in blue *light*.

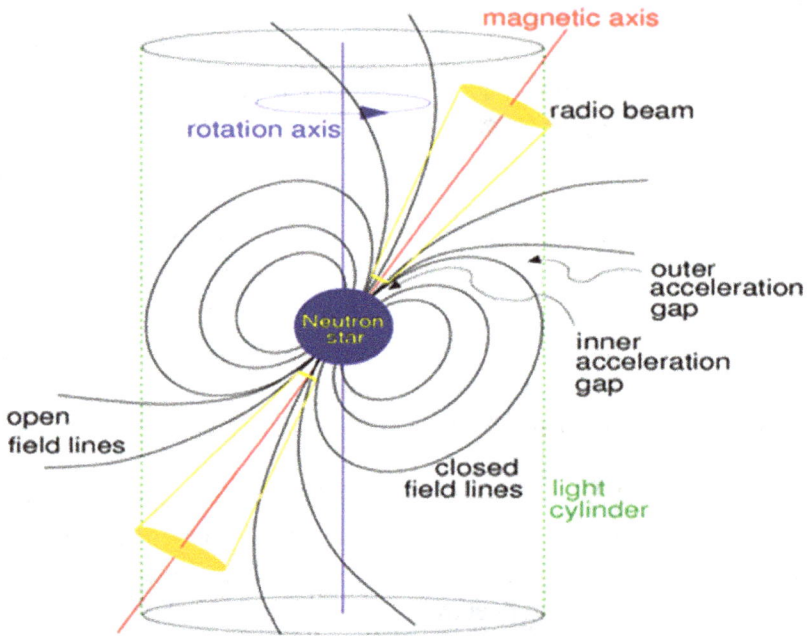

Fig. 5.3.

The Crab nebula (1054 supernova remnant) hosts a pulsar, a rapidly-spinning neutron star, and it powers the diffuse, strongly polarised bluish 'synchrotron' nebula from which the red emission-line filaments seem to emerge.

The photograph was made from plates taken on the Palomar 5m telescope in February 1956.

As (Glendenning 2000) points out there are more than 1100 pulsars discovered since the first one and in the case of Crab and Vela pulsars their periodicities 33 ms and 89 ms were decisive in identifying these objects as highly compact neutron stars. Though there have been a large number discovered in various surveys, only two seem to be in our neighbourtng galaxies LMC (PSR 0540–69) & SMC (PSR 0042–73) and the remaining in our own galaxy. Among the host of pulsars discovered the most important and most useful has been the Hulse Taylor binary pulsar (PSR 1913+16)(Hartle and Thorne 1974) where both are neutron stars and have

Fig. 5.4.

a period of about eight hours but a very large precession of the orbit ≈ 4.2 degrees per year and thus provides an excellent ground for testing general relativity. In fact, this did prove it so by confirming the existence of gravitational waves through its orbital period damping exactly as prescribed by general relativity, through the orbital energy depleting due to the emission of gravitational radiation. More on this in the chapter on gravitational waves. The fastest known pulsar was discovered in 1982 (PSR 1937+21) by the team (Backer *et al.*, 82) and has a period of 1.56 ms implying rapid rotation of the associated neutron star at the rate 650 times a second and is supposed to be the result of a spin up of the star with a low magnetic field and accreting matter slowly from a Keplerian disk, as was theorised in 1982 (Alpar *et al.*, 1982), (Bachus *et al.*, 1982) and (Radhakrishnan & Srinivasan 1982). From the available data, it looks possible that such fast rotating neutron stars are aplenty in globular clusters as more and more millisecond pulsars are being discovered.

It has also been found that few pulsars exhibit what are called 'gliches'- relatively small random and unpredictable changes in periods. (Alpar 96) whose size could vary from pulsar to pulsar and sometimes even within a pulsar. Though the birth of neutron stars is associated with Supernovae, it has not been possible to associate the remnant with a pulsar except in the case of young pulsars which could be due to the velocities with which they move away from the site of birth. As the supernova ejecta slows down after coming across the interstellar medium, it is possible that the neutron star outruns the ejecta. With all these aspects the study of supernovae and their resultant compact objects can be a very interesting and engaging research activity. Particularly after the discovery of gravitational waves coming from the coalescence of binary neutron stars in 2017 along with emissions in several wavelengths of the electromagnetic spectrum, the multi-wavelength astronomy has been a very active field of investigation today.

Chapter 6

Gravitational Collapse (Stellar Evolution Continued)

6.1 Introduction

As is known with the advent of space and satellite technology in the late 1950s, and the later part of the last century, particularly post-1960, and with advances in radio astronomy, our realisation of the universe increased almost a hundredfold larger with an awe-inspiring view of hosts of otherwise unseen objects like quasars, pulsars, X-ray binaries and active galactic nuclei. Most of these objects emit radiation in almost all known frequencies of the electromagnetic spectrum, with luminosities ranging from a few times to few thousand times the galactic emission. As summarised by G. Shields (1999), A. Sandage of the Mt. Wilson and Palomar Observatories and M. Schmidt of the Caltech took up the quest for optical identifications and measure the redshifts of radio galaxies after the identification of 3C 295 by Minkowski in 1960 with a member of a cluster of galaxies at the unprecedented redshift of 0.46. The same year, Sandage obtained a photograph of 3C 48 showing a 16th magnitude stellar object with a faint nebulosity. The spectrum of the object showed broad emission lines at unfamiliar wavelengths, and photometry showed the object to be variable and to have an excess of ultraviolet emission compared with normal stars. In 1963, Schmidt obtained the spectra of 3C 295 which showed broad emission lines at unfamiliar wavelengths which seems to have indicated that it could not be an ordinary star. Further Schmidt's keen analysis confirmed the four observed lines to be those of Hydrogen but with a large redshift as given by $z = 0.16$. This information seems to have helped him in identifying another line in the UV part of the spectrum with Mg II, which incidentally was also earlier identified in the spectrum of 3C 48. Combined with the studies made by Greenstein, and Mathews, and Oke and Hazard of the same objects it was concluded that all these special cosmic objects are extragalactic with their redshifts reflecting Hubble expansion. An accurate determination of the position of 3C 273 (Hazard *et al.*, 63) using the lunar occultation technique had revealed that the source 3C 273 has two components one of which is a jet moving away from the star. However, the more intriguing observations were related to their luminosities. Whereas the radio luminosity was comparable to that of Cyg A, the optical luminosity was about thirty times brighter than the brightest giant elliptical galaxies and the radio surface brightness was larger than for radio galaxies. Further for 3C 295, the redshift indicated a velocity of 47,400 km/sec and a distance of about 500 Mpc (for H_0 of the order 100 km/sec/Mpc). All these information put together seem to have convinced the astronomers that the nuclear region of these objects could be less than 1 kpc in diameter. The jet would be about 50 kpc away, implying a timescale greater than 105 years and a total energy radiated to be at least 10^{59} ergs. The spectrum of these objects showed broad emission lines at unfamiliar wavelengths, and photometry showed the object to be a variable having an excess of ultraviolet emission compared with normal stars. Around the same time several, other apparently star-like images coincident with radio sources were found to show strange, broad emission lines and all such cosmic sources came to be known as *quasi-stellar radio sources (QSRs), or Quasars*. In the 60s, one of the most important and most enigmatic and deeply studied aspects of astrophysics was regarding the source of energy emission of these newly discovered cosmic objects as mentioned earlier.

While special relativity had its important role in describing the physics of white dwarfs and neutron stars, the discussion of the formation and dynamics of black holes certainly calls for general relativity. As pointed out by Zeldovich and Novikov (1971), though the need for going beyond the framework of Newtonian gravitational theory was first recognised in cosmology with the advent of the Friedmann solution (1922) of Einstein's equations, the real mixing of general relativity with the rest of physics (electromagnetism, hydrodynamics, etc.,) became essential to discuss the energetics of quasars, super massive stars, X and γ ray sources. In the context of the energetics of quasars and the like, Hoyle and Fowler (1963) have pointed out that the gravitational potential energy calculated from Newtonian physics, for a spherical body of mass M and radius R is of the order GM^2/r ($M \approx 10^8 M_\odot$ and radius $\sim 10^4 R_\odot$). However, Harrison *et al.* (1965) have raised the doubt, about whether gravitational collapse, under conditions, however idealised, provides a means to convert a large fraction of the mass into energy and if it happens, what would be the final state of matter there after?

6.2 Gravitational Collapse

Landau (1932) considering this question had shown that for a model consisting of a cold, degenerate Fermi gas there exists no stable equilibrium configuration for masses greater than a certain critical mass, and had found for a mixture of electrons and nuclei the critical mass to be $\sim 1.5 M_\odot$. Oppenheimer and Volkoff (1939) considered the same problem in 1939 and pointed out that Landau's original result for a cold relativistic degenerate Fermi gas of neutrons, of $6 M_\odot$, was based purely on the Newtonian theory of gravitation, while such high mass and density would warrant general relativistic analysis. Apart from that, the gravitational effects of the neutron kinetic energy would also be important. With this in the background, they looked for a solution of Einstein's field equations for a spherical distribution of matter with massive neutron cores and came to the conclusion that 'it is very unlikely to find static neutron cores playing any part in the stellar evolution, and the question as to what could happen to stars of mass $> 1.5 M_\odot$ remains unanswered. Their point of view was, that either the equations of state used for the analysis are suspect or *the star will continue to contract indefinitely without reaching equilibrium.* They observed in their concluding remarks, that *among all spherical, nonstatic solutions one would hope to find some for which the rate of contraction and in general the time variation, becomes slower and slower so that the solutions are regarded as quasi-static and not equilibrium solutions.*

Further, for large enough mass, the central density and pressure keep growing, making the gravitational potential g_{tt} (coefficient of dt^2) become smaller and smaller, slowing down the processes for an outside observer'. This was indeed a prophetic statement as one now clearly knows that *for an external observer in the Schwarzschild geometry, the time of contraction keeps increasing as the stellar collapse continues and as r tends towards 2m, reaches infinity at r = 2m.* Before going on to discuss the final stages of gravitational collapse, it is worthwhile to look at the possible effects of general relativity with the earlier stages of contracting polytropes as discussed in the case of white dwarfs and neutron stars. In order to follow this argument it is relevant to consider the structure of geodesics in the Schwarzschild geometry and their implications.

As general relativity is a fully covariant theory and it allows one to choose any coordinate system for working out the algebra, it becomes necessary to understand the limitations that could arise with any given coordinate system. Considering the often used spherical or axisymmetric coordinates, one finds that they have an inherent lacuna, as the axis $\theta = 0$, π is not covered in the manifold since the metric becomes degenerate at these values. However, as one knows this is a removable impediment. Using the Cartesian coordinates, this degeneracy can be easily avoided. Hence one calls such a feature as a removable or coordinate singularity. But in the case of the Schwarzschild manifold, outside of the mass distribution, apart from the axis mentioned above there is the surface $r = 2m$, often called *Schwarzschild singularity*, where the coefficients of 'dt' and 'dr' behave in a singular fashion, with one going to zero and the other to infinity. Luckily this also can be removed by analytic extension of the coordinate patch. In order to understand the special features of this surface, one needs to discuss the behaviour of geodesics (trajectories of particles and light rays) through this surface. Considering the radial null geodesic, one finds, as θ and φ are constants, the equation for $ds^2 = 0$, to be

$$0 = \left(1 - \frac{2m}{r}\right)\dot{t}^2 - \left(1 - \frac{2m}{r}\right)^{-1}\dot{r}^2 \qquad (6.1)$$

(overhead dot denotes the derivative with respect to the path parameter.) Using the fact that the space-time under consideration is spherically symmetric and static, one has the energy and angular momentum of the particle to be constants of motion as expressed by (4.16). Using these in the above equation one gets $\dot{r}^2 = k^2, \Rightarrow \dot{r} = \pm k$, where $k = E/c^2$, a constant. It also follows from the above that

$$\frac{dr}{dt} = \frac{\dot{r}}{\dot{t}} = \pm c\left(1 - \frac{2m}{r}\right). \tag{6.2}$$

One can notice from this equation that if one considers a signal coming from the surface of the star to an outside observer, then the time taken for it to reach the observer would be

$$t = \int dt = \pm \int_r^\infty \left(1 - \frac{2m}{r}\right)^{-1} dr. \tag{6.3}$$

As the radius of the star reduces, the time taken keeps on increasing and as r → 2m, the time tends to infinity, showing that the signal from a star with radius equal to 2m would never reach the outside observer.
Integrating the above equation, one can get the solution

$$t = \pm(r + 2m\ln(r - 2m)) + C_1, \tag{6.4}$$

that shows one can divide the space–time *region* into separate sections, with respect to the two surfaces $r = 2m$, with (dr/dt) either positive or negative. As $dr/dt > 0$ indicates r increasing with time, the curves
$t = r + 2m \ln(r - 2m) + C_1$ represent the congruence of outgoing geodesics, while $t = -(r + 2m \ln(r - 2m)) + C_2$ represent the congruence of incoming geodesics.
Using a simple change to a new time coordinate such that the radial null geodesics become straight lines in the form $t' = t = (r + 2m \ln(r - 2m)) + C_1$, and writing the Schwarzschild metric in the coordinates (t', r, θ, φ), one finds the metric to be

$$ds^2 = (1 - 2m/r)dt'^2 - (4m/r)dt'dr - (1 + 2m/r)dr^2 - r^2 d\Omega^2, \tag{6.5}$$

This transformation given above, which apparently was first found by Eddington (1924) and later by Finkelstein (1958), gives the extension of the canonical Schwarzschild manifold to overcome the coordinate pathology at $r = 2m$, and has been very important in the developments that lead to the study of singularities in general relativity. This form of the metric is called the Eddington-Finkelstein form. As has been explained in many standard texts (Ray, 1992; Landau and Lifshitz, 1951; Misner *et al.*, 1973), the infalling particle or photon takes infinite time to reach the surface $r = 2m$ as seen in the coordinate time t, (it is also called the infinite red shift surface), while with respect to proper time s (comoving) it can pass through the surface and reach the singularity at $r = 0$ in a finite proper time as there is no catastrophe at that surface. In fact, the curvature invariant $K = R_{hijk} R^{hijk}$ on this surface is finite and equal to $48m^2/r^6$ (Prasanna, 1973). On the other hand, the outgoing geodesics both time-like and null are *trapped within the surface* $r = 2m$, as the local light cone structure is such that the null trajectories are bent inwards towards $r = 0$ for $r < 2m$, while they are bent outwards towards r going to infinity for $r > 2m$. This means, while no outgoing time–like or null geodesic can cross the surface $r = 2m$, the ones below this surface go towards the singularity $r = 0$, and those outside the surface reach the observer at infinity with a very long time delay depending upon how close to the surface they are emitted from. Further, the surface $r = 2m$ being a null surface, across the surface the nature of time–like and space–like coordinates interchange thus creating a barrier for any signal to escape from within. It is thus the Schwarzschild surface $r = 2m$ is termed as an ***event horizon or infinite red shift surface*** and spherical objects with a radius less than or equal to $2m$ are popularly known as ***black holes***. The point $r = 0$, where the scalar invariant $K = R_{hijk}R^{hijk}$ blows up to infinity is called ***a true or curvature singularity***. As only the ingoing trajectories go across the surface in finite proper time, while the outgoing ones are trapped, the Schwarzschild surface is also called a ***trapped surface* or a *one-way membrane***.

   In the same context of gravitational contraction and stellar equilibrium it is relevant to mention the critical analysis by (Chandra sekhar, 1939, 1964a, 1964b), who has shown that for any finite γ the ratio of specific heats, dynamical instability always sets in before the stellar radius reaches the so-called Buchdal limit $R > \frac{9}{8} R_s$, $R_s = 2m$, $m = GM/c^2$. Further, he established that if γ is even slightly more than 4/3, then for dynamical stability the star should satisfy the condition $R > \left[K / \left(\gamma - \frac{4}{3} \right) \right] R_s$, where the constant K depends mainly on the density distribution. The main result of this analysis has been that in the framework of general relativity, it is clear that dynamical instability through a mode of radial oscillations will intervene before the spherical gaseous mass with a polytropic equation of state can contract to the limiting radius compatible with hydrostatic equilibrium along with γ the ratio of specific heats approaching the value 4/3. This ensures that the collapsing sphere becomes unstable before reaching the Schwarzschild limit. It is important to note that in this analysis, Chandrasekhar had made an estimate of the radius, for the instability to set in for gaseous configurations. For a mass $10^8 M_\odot$ and polytropic index 3, the radius turns out to be, $R = \frac{1.1245}{\left(\gamma - \frac{4}{3}\right)} R_s \cong 4.7 \times 10^{17} cms$. This value of mass and radius matched well with the estimates made for quasi-stellar objects. (RSS 65).

6.3 Non-static Configurations

Having seen that the spherically symmetric static configurations cannot yield stable solution for the late stages of stellar evolution, Oppenheimer and Snyder (1939) continuing the earlier discussion Oppenheimer and Volkoff (1939) studied the gravitational contraction for non-static spherical distribution. As mentioned earlier to describe the possible scenario, one is required to consider non-static distribution of fluid with a possible scenario and thus they considered the case of a spherical dust ball (pressure less incoherent matter) in a system of commoving coordinates (observer moving with the collapsing sphere).

Most importantly they showed that for the comoving observer, the total time of collapse is finite, whereas an observer at infinity would find the object shrinking to its Schwarzschild radius only asymptotically. Though this result in fact agrees with the analysis of geodesics as mentioned above in principle, the assumption of zero pressure distribution appears rather unphysical. Several others have tried considering the relativistic hydrodynamics of perfect fluid distribution with nonzero pressure using the well-known Tolman–Oppenheimer–Volkoff (TOV) equations of hydrostatic equilibrium and several different equations of state (including possible nuclear forces) but have concluded that while finding an analytic solution to general relativistic equations with such constraints is difficult, only numerical solutions may be possible (Misner and Sharp, 1964; May and White, 1966).

In such a situation it was also necessary to consider the probability whether some physical process or break in symmetry could stop the collapse to singularity. As Penrose (1965) points out 'for a sufficiently enough positive mass there is no final equilibrium state and as sufficient thermal energy gets radiated away the body continues to contract and reaches the physical singularity at $r = 0$. The presence of a singularity presents a serious problem for any complete discussion of the physics of the interior region for collapsing stars. However, with a detailed discussion of the geodesics, Penrose concludes that any deviation from spherical symmetry also cannot prevent the occurrence of space-time singularity if the matter inside is always of positive energy and allows the development of a trapped surface. This shows that stars evolving into *black holes is a natural phenomenon which could possibly allow the existence of a large number of cosmic sources with very high energy output, particularly among the Quasars and the Active Galactic Nuclei.*

There seems to be strong evidence for two types of black holes: stellar black holes with masses of a dozen or so suns, and supermassive black holes with masses of many millions of suns. Stellar black holes could have formed as a natural consequence of the evolution of massive stars. The origin of supermassive black holes is still a mystery. They are found only in the centers of galaxies. It is still not known whether they formed in the initial collapse of the gas cloud that formed the galaxy, or from the merger of a centrally located cluster of black holes.

Before moving on to the discussion of the energetics associated with black holes, it is important to consider the role of rotation in the context of black hole formation. In fact, at one stage, there was a naïve thinking about whether rotation could prevent continuous collapse with the centrifugal force overcoming gravity. However, this kind of thinking was resolved with the discovery of an exact solution for the exterior of a rotating star within the framework of general relativity by R. Kerr (1963) which was found to admit all the required features of a continuous collapse and indeed got to be known as the solution for describing the *field of a rotating black hole.*

In this context, it is useful to consider the earlier work on rotation in general relativity. In 1918, Thirring and Lense looked for the effect of rotation on space by considering the effects on test masses and inertial frames inside (near the center) and in the exterior of a rotating mass shell, using a weak field approximation of general relativistic field equations. A useful discussion of this may be found in O'hanian and Ruffini (1976) who express the final form of the solution to Einstein's equations in the weak field approximation as expressed by the metric

$$ds^2 = \left(1 - \frac{2\Phi}{c^2}\right)dt^2 - \left(1 + \frac{2\Phi}{c^2}\right)(dx^2 + dy^2 + dz^2) + \Omega(ydx - xdy)dt. \quad (6.6)$$

The off-diagonal terms are interpreted as due to the rotation of the frame of reference that arises due to the Coriolis force as known in Newtonian mechanics. Here $\Omega = 4GM\omega/3Rc^2$, ω being the angular velocity and $\Phi = GM/r$ representing the Newtonian potential.

Writing in terms of spherical polar coordinates, this takes the form

$$ds^2 = \left(1 - \frac{2m}{r}\right)dt^2 - \left(1 - \frac{2m}{r}\right)^{-1}[dr^2 + r^2(d\theta^2 + sin^2\theta d\varphi^2)]$$

$$- \frac{4am}{r}sin^2\theta d\varphi dt \qquad (6.7)$$

There appears quite a lot of discussion on the work of Lense and Thirring, for a historical account of which one may refer to (Pfister 1995) or (Mashoon *et al.*, 1984). One of the main results of these discussions is the interpretation of the Coriolis force which appears outside but very close to the rotating body and is normally referred to as the *dragging of inertial frames*. It is known that in electromagnetics, the electric charge on a static sphere produces only an electric field but on a rotating sphere, it produces also a magnetic field. Thus in similarity, the dragging of inertial frames by a rotating body is called *gravimagnetic effect*. It is interesting to learn that the effect can be used as a test for general relativity by measuring the precession of gyroscopes in earth's orbit.

6.4 Kerr Solution and Its Features

The Kerr solution (1963) of Einstein's equations for empty space ($R_{ij} = 0$), which defines the external field of a rotating star is expressed in Cartesian coordinates as given by the axisymmetric, stationary metric:

$$ds^2 = dt^2 - (dx^2 + dy^2 + dz^2) - [2mr^3/(r^4 + a^2 z^2)]dw^2, \qquad (6.8)$$

where $dw^2 = [dt + \frac{r}{(a^2 + r^2)}(xdx + ydy) + \frac{a}{(a^2 + r^2)}(ydx - xdy) + \frac{z}{r}dz]^2$. It is easy to check that by simplifying and dropping the terms of order a^2 and higher powers in a one gets the metric

$$ds^2 = \left(1 - \frac{2m}{r}\right)dt^2 - \left(1 + \frac{2m}{r}\right)(dx^2 + dy^2 + dz^2) - \frac{4am}{r}(ydx - xdy)dt \quad (6.9)$$

which is exactly the Lense–Thirring metric for a rotating disk discussed above. Using spherical polar coordinates, Boyer and Lindquist (1967) rewrote the Kerr metric in what is called Boyer–Lindquist coordinates

$$ds^2 = -\left(1 - \frac{2mr}{\Sigma}\right)dt^2 - \frac{4mra}{\Sigma}\sin^2\theta \, dt \, d\varphi + \frac{\Sigma}{\Delta}dr^2 + \Sigma \, d\theta^2$$
$$+ \frac{B}{\Sigma}\sin^2\theta \, d\varphi^2 \qquad (6.10)$$

$$\Delta = (r^2 - 2mr + a^2), \; \Sigma = (r^2 + a^2\cos^2\theta), \; B = (r^2 + a^2)^2 - \Delta a^2\sin^2\theta.$$

It is easy to check that putting $a = 0$, in this metric results in the already discussed Schwarzschild solution for the external field of a static spherical star. It is worth looking at the limit of slow rotation by expanding the various factors and neglecting the terms of the order a^2 and higher powers in a. The metric so obtained is given by

$$ds^2 = -\left(1 - \frac{2m}{r}\right)dt^2 - \frac{4ma}{r}\sin^2\theta \, dt \, d\varphi + \left(1 - \frac{2m}{r}\right)^{-1}dr^2 + r^2 d\Omega^2 \quad (6.11)$$

which again matches with the Lense-Thirring metric for a slowly rotating body in the weak field limit of Einstein's equations. This further confirms the *interpretation of Kerr metric as the exact solution for an axi-symmetric stationary system representing a rotating star, in fact, a rotating black hole.*

6.5 Special Features of Kerr Space-Time

Unlike Schwarzschild static geometry, Kerr geometry exhibits more dynamical features due to the richer functional forms of the metric potentials, $g_{tt} = \left(1 - \frac{2mr}{r^2 + a^2 \cos^2 \theta}\right)$ and $g_{rr} = -\frac{(r^2 + a^2 \cos^2 \theta)}{(1 - 2mr + a^2)}$. First thing one can notice is that both $g_{tt} = 0$ and $g_{rr} = \infty$ give different surfaces and thus the space-time slicing would be different than for static black holes. The four different surfaces are defined as follows:

$$g_{tt} = 0, \text{ implies } r^2 - 2mr + a^2 \cos^2 \theta = 0, =>$$
$$r_0^{\pm} = m \pm \sqrt{(m^2 - a^2 \cos^2 \theta)}, \tag{6.12}$$

and

$$g_{rr} = \infty, \text{ implies } r^2 - 2mr + a^2 = 0, =>$$
$$r_{\infty}^{\pm} = m \pm \sqrt{(m^2 - a^2)} \tag{6.13}$$

that can be depicted as in the figure: ($r_0^{\pm}$ are in red and $r_{\infty}^{\pm}$ are in black).

It is clear that these surfaces are defined (real) only for the case $m \geq a$ and when $m = a$ it is referred to as the extremal case. It may be noted that for the case $a = 0$, both r_0 and r_{∞} are equal to $2m$ as it should be for the Schwarzschild case. It may also be seen that the two radii r_0 and r_{∞} coincide at the poles $\theta = 0, \pi$, whereas they are separate at the equator $\theta = \pi/2$,

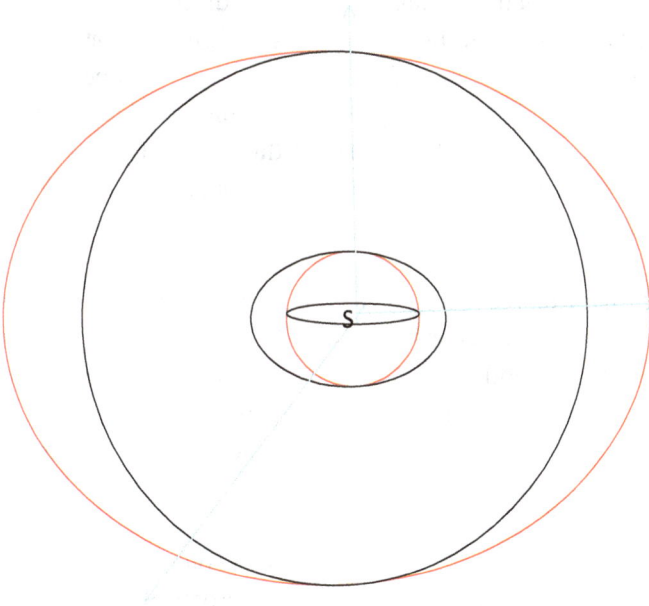

Fig. 6.1.

showing that there is a bulge at the equator. The incoming particle thus can enter the region between r_0^+ and r_∞^+ and escape out without getting sucked in by the event horizon. Thus the incoming particle can move back after crossing r_0^+ but gets trapped if it crosses r_∞^+. The surface r_0^+ is also called the *ergosurface* and the region in Fig. 6.1 between r_∞^+ and r_0^+, is called the *ergosphere*. This nomenclature comes from the fact that as a particle entering this region can escape back with higher energy than it came in with, due to a process called the Penrose process of energy extraction (Penrose, 1969). The extra energy that the outgoing particle carries comes from the rotational kinetic energy of the black hole.

It was quite clear by the mid-sixties that the three main stages of a star's evolution end in three possible end results As summarised by Glendenning (2000), the Oppenheimer-Volkoff analysis provides the equilibrium configuration at each central density corresponding to an equation of state for cold matter. As has been discussed by Wheeler *et al.* (1965) for a polytropic equation of state there are a number of stellar sequences satisfying the stability condition $dM/d\rho_c > 0$, but however, beyond the neutron star density, all are unstable to vibrational modes. The squared frequency of the fundamental mode as discussed by Chandrasekhar (1964a) turns negative around the density 10^9 gms/cc that terminates the white dwarf configurations and interestingly the fundamental mode gets stable again around the density $1.5 \ 10^{14}$ gms/cc which is the density requirement for stable neutron stars. At this stage, it is safe to conclude that a collapsing core of a star could end up as one of the following three possibilities.

(1) If the mass of the collapsing core is less than the 'Chandrasekhar limit' of 1.4 solar masses, then the electron degeneracy pressure will prevent further collapse and the star ends up as a **white dwarf**.

(2) If the mass of the collapsing core is about **2 to 4** solar masses then the neutron degeneracy pressure stops further collapse and the star ends up as a **neutron star**.

(3) If the collapsing core is of mass greater than about **5 to 8** solar masses, then no other force can come into play to counter gravity and the collapse continues on till the configuration crosses its *event horizon* and end up as a **black hole**.

Chapter 7

Black Hole Energetics and Accretion Physics

7.1 Introduction

If black holes by definition are non-emitting objects from inside of which no radiation can ever reach the observer at infinity, how could one observe them and how can one study them? This doubt must have arisen in one's mind for which the answer goes as follows. As black holes are extremely compact massive objects, their gravitational potential is very high. This would make them sink for any type of matter that nears it and particularly matter from a slowly evolving companion star or from interstellar matter. Such a process of matter getting closer to a star is referred to as the process of accretion. The phenomenon of accretion onto compact objects has been a subject of study for a long time.

The accretion efficiency varies between compact objects and is given by $\eta = GM/Rc^2$, where M is the mass and R is the radius of the compact object. G and c are universal constants. For a white dwarf, $\eta = 1.2 \times 10^{-4}$ and for a neutron star $\eta = 0.15$. If $\dot{M}$ is the mass accretion rate then the luminosity due to accretion is given by $L = \eta \dot{M} c^2$. In the case of black holes, as there is no hard surface on which the matter can accrete, the efficiency is calculated as the amount of rest mass energy liberated from the last stable circular orbit of the particle accreting onto the black hole. This can be calculated using the effective potential as defined earlier which for a Schwarzschild black hole is given by

$$V_{eff}^2 = \left(1 - \frac{2m}{r}\right)\left(1 + \frac{l^2}{r^2}\right),$$ (7.1)

where l is the specific angular momentum of the particle given by

$$l^2 = mr \Big/ \left(1 - \frac{3m}{r}\right).$$

As the last stable orbit for a free particle in Schwarzschild geometry is at $r = 6m$, one finds the efficiency η to be $\eta = 1 - V_{eff} \approx 0.057$. In the case of a rotating black hole as the last stable orbit in Kerr geometry is much more closer to the horizon, the efficiency turns out to be ≈ 0.42. In fact, this is the highest for any compact object.

The phenomenon of accretion was initially recognised in the study of cosmogony (the study of planet formation in the solar system) (Treves *et al.*, 1989) but the real impetus came from the work of Hoyle and Lyttleton (1939) who examined the possible change in luminosity of a sun-like star in its passage through the interstellar medium. Though theirs was the first derivation of the accretion rate for a star moving through a cold gas, Bondi (1952) was the first to bring in hydrodynamics into the description. He considered the infalling gas onto a static gravitating body through a polytropic equation of state and obtained a full analytic solution for the fluid flow and also calculated the accretion rate. However, serious study of accretion physics came much later with the discovery of quasars and X-ray sources with the studies of Salpeter (1964) and Zeldovich (1964), trying to account for the enormous luminosities of these objects through accretion onto massive collapsed stars. It was realised that when the accreting matter has angular momentum, its accretion required the removal of angular momentum which in turn required dissipative forces like viscosity. These in turn changed the scenario with the study of the formation of accretion disks around the compact objects. Pringle and Rees (1972) and Shakura and Sunyaev (1973) have made a detailed study of such accretion disk models. Further, Prenderghast and Burbidge (1968) used the disk model to explain the phenomenon of cataclysmic variables while Lynden Bell (1969) used a similar model to account for emission from the center of the galaxy to be due to accretion disk around a massive black hole at the centre.

7.2 Physics of Accretion

The energy extracted through accretion comes out in the form of electro-magnetic radiation, thus increasing the luminosity of the accreting body. If all the kinetic energy of the infalling matter is converted to radiation at the surface of the accreting body, one can then define the accretion luminosity $L = GM\dot{M}/R$, which may be approximately expressed as

$$2 \times 10^{32} \left(\frac{GM}{Rc^2} \right) \left(\frac{\dot{M}}{M_{sun}} \right) \left(\frac{10^4}{T} \right)^{3/2} N \text{ ergs/s}, \tag{7.2}$$

where $\dot{M}$ is the steady accretion rate, T is the temperature, and N is the number density of gas near the accreting source (Zeldovic and Novikov (1972)).

The important question that arises at this point is, can a body keep on increasing its luminosity by continued accretion? Unfortunately not! The limit comes from a very simple fact of physics. As the energy from accretion is converted to radiation, on a typical gas particle near the accreting source, there are two forces acting–gravity pulling it inwards and radiation pressure pushing it out. Assuming the accreting gas to be composed of fully ionised hydrogen, the radiation pressure mainly comes through Thomson scattering such that if S is the radiant energy flux expressed in units of (ergs $s^{-1} cm^{-2}$), then the outward radial force on each electron as given by the rate at which it absorbs momentum, is $\sigma_T S/c$, σ_T being the Thomson cross-section, which is equal to 6.7×10^{-25} cm^2. Due to the electrostatic Coulomb interaction, the electron generally drags the proton (ion) and thus the couple gets pushed away against the total gravitational pull as given by $GM(m_e + m_p)/r^2$ at any distance r. If L is the luminosity of the source then $S = L/4\pi r^2$, and the net inward force on the pair would be given by

$$F = (GMm_p - L\sigma_T/4\pi c)/r^2. \tag{7.3}$$

Thus, the limiting luminosity is attained when the force is zero (the two opposing forces being equal) which is called the *Eddington luminosity* and is given by

$$L_{edd} = (4\pi Gm_p c/\sigma_T) \cong 1.3 \times 10^{39} M/M_{sun} \quad \text{ergs/s}. \tag{7.4}$$

The accretion would stop once this critical luminosity is reached. This in turn gives the critical accretion rate to be

$$\dot{M}_c = \frac{L_{edd}}{c^2} = \frac{4\pi GMm_p}{c\sigma_T}. \tag{7.5}$$

It is to be noted that the above value $\dot{M}_c$ was obtained for steady spherical accretion, but there can be situations when the accretion rate could exceed $\dot{M}_c$

The scenario presented above is relevant only for objects like neutron stars and white dwarfs, which have a surface on to which the accreting matter can impinge upon and lose its kinetic energy. On the other hand, if the compact object is a black hole, there will be no hard surface and the matter just disappears below the event horizon, through which a large part of the rest mass energy would get lost. However, the difference is that for black holes, the capture radius is much smaller than that for a neutron star, as well as the mass of the black hole could be much higher, leading to higher efficiency. (An interesting but limited aspect of this process of accretion is that, while there exists no stationary solution for steady accretion in the case of neutron stars and white dwarfs for $\dot{M} > \dot{M}_c/\eta$, in the case of black holes η can be reduced so that the black hole can swallow any amount of accreting matter but giving out only a limited amount of energy.) With this process of energy emission, what type of electromagnetic radiation could one find? One can make some useful order of magnitude estimates of the emitted spectra as follows.

If v is the frequency of a typical photon emitted by radiation of temperature T_{rad}, then one has $T_{rad} = hv/k$ (h and k being Planck and Boltzmann constants respectively). If T_b is the black body temperature corresponding to the emission from a source of radius R and accretion luminosity L, then $T_b = (L/4\pi R^2\sigma)^{1/4}$. If the gravitational potential energy of the accreted material is entirely converted to thermal energy, for each electron–proton pair accreted, the potential energy released is of the order GMm_p/r. As the thermal energy is given by the expression $2 \times \frac{3}{2}kT_{th}$, one finds the temperature associated with this energy is $T_{th} \cong GMm_p/3kR$. If the accretion flow is optically thick, as the radiation has to reach thermal equilibrium with the accreting material before escaping, one should have $T_{rad} = T_b$. On the other hand, if the accretion column is optically thin, as the accretion energy gets converted to radiation without any further interaction, $T_{rad} = T_{th}$. Thus, in general, the radiation temperature lies between the limiting temperatures, $T_b \leq T_{rad} \leq T_{th}$. If one considers accretion onto a neutron star ($M \simeq 1\ M_{sun}$ and $R \sim 10$ kms,) then $T_{th} \simeq 5.5 \times 10^{11}\ K^0$ giving $kT_{th} \sim 50$ Mev. Considering the Eddington luminosity as the limit, with $L_{acc} \sim 10^{32}$ ergs/s one finds $T_b \simeq 10^7$ K degrees, giving $kT_b \sim 1$ Kev. Thus one can get the range of energy emitted to be 1 Kev $\leq h\nu \leq 50$ Mev, indicating the emission of medium to high energy X-rays. Scaling down the luminosity appropriately with increased radius for the case of white dwarfs, one can expect the emission to be in the range, 6 ev $\leq h\nu \leq 100$ Kev, giving optical to UV or low energy X-rays (Frank *et al.*, 1985).

As mentioned in the beginning, there are two modes of accretion:

1. Spherical accretion where the infalling particles (having no angular momentum) follow radial geodesics of the background geometry and fall inwards onto the compact star.
2. Disk accretion where the infalling particles having angular momentum move along the Keplerian orbits around the compact object and will have to lose their angular momentum to spiral slowly inwards towards the compact object.

The earliest known theoretical treatment of the phenomena of accretion was due to Hoyle and Lyttleton (1939). They had considered the possibility of Sun accreting inter-stellar matter during its motion around the galaxy. When a star of mass M moves with a velocity v in a cold gas of density ρ, the particles of gas follow Keplerian orbits in the field of the star. When their transverse momentum gets dissipated through collisions at a distance r from the star, they are captured by the star if their parallel velocity is lower than the escape velocity of the star. This distance r where this can occur is called the capture radius and is denoted by the symbol r_c. This will initiate a slow process of accretion by the star with the rate

$$\dot{M} \approx \pi \rho v\, r_c^2 = 4\pi \rho G^2 M^2 / v^3.$$

If the scenario was to be slightly different with the star being at rest and the fluid in motion, having a strong attraction from the star upto the region wherein the sound velocity in the fluid is less than the escape velocity v then the capture radius $r_c = 2GM/v_s^2$, and $\dot{M} \approx \pi \rho v_s r_c^2$. This in turn gives for a star moving in a fluid having sound speed $v_s \propto \sqrt{\left(\frac{dp}{dp}\right)}$ the accretion rate (Treves *et al.*, 1989) given by

$$\dot{M} = [4\dot{\pi}(GM)^2/(v^2 + v_s^2)^{3/2}]. \tag{7.6}$$

Bondi (1952) treated the dynamics of spherical accretion and the relevant equations in the Newtonian dynamics for the stationary state are as follows:

the mass conservation law: $\dot{M} = 4\pi r^2 v \rho,$ \hfill (7.7)

the Euler equation $v\dfrac{dv}{dr} = -\left(\dfrac{1}{\rho}\dfrac{dp}{dr} + \dfrac{GM}{r^2}\right)$ \hfill (7.8)

and the equation of state $p = p(\rho)$, which for the polytropic case is $p \propto \rho^\gamma$, $1 \le \gamma \le 5/3$. Put together these equations give the condition,

$$\frac{v^2}{2} + \frac{1}{(\gamma - 1)}v_s^2 - \frac{GM}{r} = \text{constant}, \quad v = \frac{\dot{M}}{4\pi\rho_\infty r^2}\left(\frac{v_{s\infty}}{v_s}\right)^{2/(\gamma-1)}, \tag{7.9}$$

$\rho_\infty, v_{s\infty}$ being the values of fluid density and the sound speed far away from the accreting source. One can plot for different values of r in the (v, v_s) plane.

The first of the above two equations gives a set of ellipses whereas the second gives a set of hyperbole. These curves intersect at two points, one corresponding to sub-sonic flow and the other to super-sonic flow as depicted in the Fig. 7.1 (ZN 72). If the fluid velocity $v = 0$ at infinity, then it may be seen that the constant in (1) has to be $v_{s\infty}^2/(\gamma - 1)$. Thus for $v = v_s$, the sonic point, $v_s^2(r_s) = \frac{2v_{s\infty}^2}{5 - 3\gamma}$. which shows that the sonic point occurs at $r_s = \frac{(5-3\gamma)GM}{4v_{s\infty}^2}$ where the accretion rate turns out to be

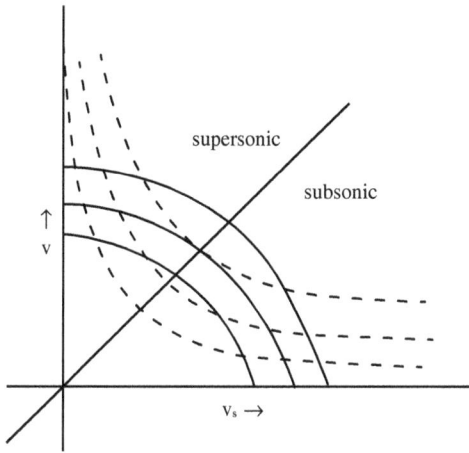

Fig. 7.1.

$$\dot{M} = (\pi GM^2 \rho_\infty / v_{s\infty}^3)(2/(5-3\gamma))^{(5-3\gamma)/(2(\gamma-1))}. \tag{7.10}$$

For monatomic non-relativistic gas with $\gamma = 5/3$, the rate is $\dot{M} = (\pi GM^2 \rho_\infty / v_{s\infty}^3)$. (Applying the formula $\lim_{x \to 0}(1/x)^x = 1$). Another point to note is that as for the value $\gamma = 5/3$, the sonic point occurs at $r = 0$, and the flow is everywhere subsonic. This situation needs careful considera-tion as the interstellar gas at a typical temperature of the order 10^4 K, will be partially ionised and thus when compressed use up some energy which could make the value of γ different from the value 5/3, and thus generally assumed to be ~1.4. Taking a look at the plot (v, v_s), one can find that for a given value of $\dot{M}$ and ρ_∞, the curves may not intersect for any r. In such a situation, one of the possibilities is that the flow is subsonic at infinity and the curves touch one another only tangentially and the flow gets supersonic after crossing r_s. The other possibility is when the flow is sub-sonic everywhere and the curves do not touch each other even tangentially for any r in which case accretion would be possible only when the pressure near the stellar surface is sufficiently high. If the flow is supersonic near the stellar surface, accretion could become feasible only if the scale height $H = \frac{N_0 kTR^2}{GM\mu}$ (N_0 is the Avogadro number and μ is the mean molecular weight) is such that $H + R < r_s$. As the transonic flow near the star is almost with free fall velocity $\sim \sqrt{(2GM/r)}$, the temperature of the gas has to satisfy $T \ll 10^4 \left(\frac{M/M_{sun}}{R/R_{sun}} \right)$. Zeldovich and Novikov (1972) point out that the gas density remains approximately constant from far away (r approaching infinity) down to the critical radius r_c where the gravitational potential is of the order $v_{s\infty}^2$, giving $r_c = \Gamma GM/v_{s\infty}^2$, with $\Gamma(\gamma) = \frac{1}{2}$ $[2/(5-3\gamma)]^{(5-3\gamma)/2(\gamma-1)}$. For different values of the adiabatic index, one has $\Gamma\left(\frac{5}{3}\right) = \frac{1}{2}$ and $\Gamma\left(\frac{4}{3}\right) = 1$. As the transition from subsonic to supersonic occurs at the radius $r_s = \frac{5-3\gamma}{4} \left(\frac{GM}{v_{s\infty}^2} \right)$, the velocity at this point is $u_s = v_s = v_{s\infty} \left(\frac{2}{5-3\gamma} \right)^{(1/2)}$. For the polytropic equation of state, the density at r_s is given by $\rho(r_s) = \rho_\infty \left(\frac{v_s(r_s)}{v_{s\infty}} \right)^{2/(\gamma-1)}$. Using the notations as in (ZN72) and introducing the parameter $\alpha = \left(\frac{\pi}{4\gamma^{3/2}} \right) \left(\frac{2}{5-3\gamma} \right)^{(5-3\gamma)/2(\gamma-1)}$, and the perfect gas equation of state $p = nkT$ at infinity, one can rewrite the accretion rate in the following equivalent forms:

$$\dot{M} = 4\gamma^{3/2}\alpha GM^2 \rho_\infty / v_{s\infty}^3 \text{ or } \dot{M} = \alpha R_g^2 c \rho_\infty (m_p c^2 / kT_\infty)^{3/2}; \qquad (7.11)$$

with $R_g = 2GM/c^2$. The constant α for different γ is given by, $\alpha (1) = 1.5$, $\alpha (1.4) = 1.2$, $\alpha (5/3) = 0.3$. An important thing to follow is the fact that for non-interacting particles (instead of gas), the accretion rate would be $\dot{M}_p = 4\pi R_g^2 \rho_\infty c^2 / v_\infty$, v_∞ being the free fall velocity for the particle. Comparing this with the accretion rate for gas (hydrodynamic approximation) as given by $\dot{M}_g \cong \pi R_g^2 \rho_\infty c^4 / v_{s\infty}^3$, one finds that the hydrodynamical accretion rate is about $(c/v_\infty)^2 \approx 10^9$ times larger than the rate for particle accretion. This could be mainly due to the fact that as collision dominates in a gas, it limits the growth of tangential velocity during infall and allows the radial velocity to grow. For more details on this discussion, one may refer to (ZN72). As pointed out by Shapiro and Teukolsky (1983), in general, the spherical accretion of interstellar gas onto stellar mass black holes is not a very efficient mechanism for radiation emission as compared to that of accretion onto neutron stars which have efficiency ~0.1 or the disk accretion onto black holes with efficiency ~0.05–0.42. The analysis requires a self-consistent treatment of the hydrodynamical equations of motion and of radiative transfer. Performing the required analysis, they (Shapiro and Teukolsky, 1983) demonstrate that the results depend very much on the regime and boundary conditions far from the black hole which may be very uncertain. It is with this background that one proposes to take a glimpse at the second scenario of accretion, *viz.* accretion disks.

In fact considering the natural configurations where one studies accretion, it is not a very sound idea to assume radial inflow of matter on to compact objects as matter in all forms do have angular momentum and disk accretion seems to be a greater possibility for radiation emission mechanisms. It is in fact elementary knowledge that a particle having angular momentum would settle down in an orbit around the central object with the star's gravitational pull being counter balanced by the centrifugal force of the moving particle. The analysis of the 'effective potential' of a particle in orbit clearly shows that for any angular momentum distribution, there exists a corresponding 'centrifugal barrier' which controls the behaviour of the particle depending upon its energy. While spherical accretion with Newtonian formulation could still be an applicable process for the case of white dwarfs and neutron stars, for black holes one requires the gravity to be treated in a general relativistic formulation. The basics of this theory were already discussed in an earlier chapter where it was shown that the equations of motion for a particle in the gravitational field are given by the geodesics of the space-time underlying the given geometry. However, before setting on to discuss the dynamics of accretion with disks, it is necessary to look into other interactions that may prevail in the cosmic scenario, with material distributions and associated fields. As accretion involves the flow of matter from the surrounding space or from an associated binary star, it is useful to briefly go through aspects related to hydrodynamics and plasma physics.

Chapter 8

Fluid Dynamical and Magneto Hydrodynamical Concepts

As accretion involves the study of matter from the surroundings onto the compact object, one needs to analyse its behaviour according to the dynamical laws that govern the system. As mentioned earlier, the inflowing matter could be a neutral gas or a charged fluid and studying their behaviour involves ordinary fluid mechanics or magnetohydrodynamics (plasmas with special characteristics) both in the Newtonian regime and in the relativistic regime. The basic equations of fluid mechanics can be found in many standard texts (Goldstein, 1960).

We consider here only a few special features for general fluids and viscous fluids. The general motion of any fluid under an external force can be considered as a deformation that need not be steady as its velocity field could have both divergence $(\nabla \cdot v)$ and curl $(\nabla \times v)$ of the velocity nonzero. In other words, if v^a are the components of velocity along the three axes, x, y, z, at a point $P(x^a)$, then at a neighbouring point Q with coordinates $x^a + \delta x^a$, the relative velocities are given by

$$\delta v^a = (\partial v^a/\partial x^b)\, \delta x^b = (e^{ab} - \xi^{ab})\, \delta x_b/2, \qquad (8.1)$$

and

$$e^{ab} = (\partial v^b/\partial x^a + \partial v^a/\partial x^b), \text{ and } \xi^{ab} = (\partial v^b/\partial x^a - \partial v^a/\partial x^b). \qquad (8.2)$$

where the symmetric tensor e^{ab} is associated with extension along the field line while the antisymmetric tensor ξ^{ab} represents rotation with the vorticity axial vector ω defined through the anti-symmetric three tensor ε_{abc},

$$\omega_a = \varepsilon_{abc}\, \xi^{bc} = (\nabla \times v)_a. \tag{8.3}$$

The relative motion thus consists of a motion in the direction perpendicular to the surface, $\Sigma = e_{ab}\delta x^a \delta x^b$, called the rate of strain quadric, and a rotation, with the angular velocity of the fluid element being equal to one-half of the vorticity. The principal axes of this quadric, which are indeed the principal axes of the tensor e_{ab}, are also known as principal axes of the rate of strain, and this part of the motion is called *pure strain* (Goldstein, 1960). If the components of angular velocity are all zero, the motion is said to be *irrotational*.

When the fluid motion is irrotational, one has, by definition, $\nabla \times v = 0$, which means one can express the velocity as the gradient of a potential φ, $v = -\nabla\varphi$. If, further, the fluid is incompressible, meaning (ρ = constant), then the continuity equation will lead to the condition $\nabla \cdot v = 0$. Equivalently, one has the condition $\nabla^2 \varphi = 0$. Such a flow is also called a potential flow and φ the velocity potential. The equation $\nabla^2 \varphi = 0$ is the *Laplace equation*.

8.1 Viscous Fluids

As there could always be frictional forces between different layers of the fluid, like viscosity or non-uniform rotation of the fluid, often in naturally occurring fluids, the transverse stresses would not be zero. It is the relative motion of fluid particles between different layers that gives rise to transverse stresses. This had led Newton to propose that the shear force that exists between layers of fluid should be proportional to the velocity gradient in the perpendicular direction. With this, if one now considers the stress tensor S_{ab}, it can be decomposed as

$$S_{ab} = p\delta_{ab} + \pi_{ab}, \tag{8.4}$$

where p represents the isotropic pressure along the three axes, also called the principal stress (δ_{ab}, the Kronecker tensor) and π_{ab} is the viscous stress tensor to be defined. From the elongation tensor e_{ab}, defining the principal rates of extension, $e_{aa} = 2e_a$, for $(a = 1\ldots3)$, one can define the scalar of expansion (or extension)

$$\theta = (e_1 + e_2 + e_3)/2 = v^a_{,a} = \sum \frac{\partial v_a}{\partial x^a} = \nabla \cdot v. \tag{8.5}$$

With these, the principal rates of viscous stresses, π_a, $(a = 1, 2, 3)$ are expressed as

$$\pi_a = \left(\varsigma - \frac{2}{3}\mu\right)\theta + \mu e_a \tag{8.6}$$

or equivalently in the tensorial form

$$\pi_{ab} = \varsigma\theta\delta_{ab} + \mu\left(e_{ab} - \frac{2}{3}\theta\delta_{ab}\right) \tag{8.7}$$

in which ς and μ represent the coefficients of *bulk* and *shear* viscosity, respectively. With this, the equations of motion for general fluids take the form

$$\rho\frac{dv_i}{dt} = \rho F_i - \frac{\partial p}{\partial x^i} + \frac{\partial \pi_{ij}}{\partial x^j}. \tag{8.8}$$

which is generally called the *Navier–Stokes equation*. Along with the equations of continuity and the energy conservation equation, this forms the complete set of hydrodynamical equations for general fluids acted upon both by internal and external forces.

Sometimes this equation is also written in the form

$$\rho\frac{dv_i}{dt} = \rho F_i - \frac{\partial p_{ij}}{\partial x^j} \tag{8.9}$$

$$P_{ij} = p\delta_{ij} - \pi_{ij}. \tag{8.10}$$

8.2 Rayleigh Criterion

An important concept associated with the rotational flow of viscous fluids concerns the transport of angular momentum, as the internal stresses between different layers of the fluid lead to differential rotation. In such a situation, if a ring of fluid from radius r_1 having velocity v_1 gets interchanged with a ring at radius r_2, having velocity v_2, the conservation of angular momentum requires the displaced fluid element to have a new velocity $(r_1/r_2)v_1$. Thus the ring which had a centripetal acceleration of (v_2^2/r_2) will now have a new acceleration $[(r_1v_1)^2/r_2^3]$ to remain in the new position. If the fluid system has to be stable, then the fluid element will have to be pushed back to its original position, which requires the condition

$$\frac{r_1^2 v_1^2}{r_2^3} < \frac{v_2^2}{r_2}, \Rightarrow (r_1^2 \Omega_1^2)^2 < (r_2^2 \Omega_2^2)^2, \tag{8.11}$$

where Ω represents the angular velocity. This condition is normally expressed in the statement: *For the stability of a rotating fluid configuration, the angular momentum has to increase outwards from the axis of rotation,* and is called *the Rayleigh criterion for stability,* mathematically expressed as $\frac{d}{dr}[r^2\Omega]^2 > 0$. (Rai Choudhuri, 1988; Rayleigh, 1917).

8.3 Magneto-hydrodynamics

As mentioned in the introduction, when a magnetic field is present and the matter is in a plasma state then one in principle should use the methods of kinetic theory (statistical mechanics). However, this gets too complicated and particularly as there is no developed theory of general relativistic kinetic theory one resorts to the methods of magneto-hydrodynamics as sketched below. This is sufficient as it treats fluids as infinitely conducting in the presence of electromagnetic fields for the following reasons. The usual characteristic lengths, such as mean free paths of particles and Debye shielding distances, are of molecular dimensions, and thus not significant for the fluid dynamics and allow the charge separation to be negligible. In principle, the application of MHD approximation for plasmas requires the normal length scales to be larger than the Debye length and the time scales to be larger than the inverse of plasma frequency. The governing equations for a system of charged particles in this approximation may be written as the equation of continuity, $\frac{d\rho}{dt} + \rho(\nabla.v) = 0$, and the modified Navier–Stokes equation as,

$$\frac{dv}{dt} = F - \frac{1}{\rho}\nabla p + \frac{\bar{j} \times \bar{B}}{c_\rho} + \frac{\mu}{\rho}\nabla^2 v + \frac{1}{\rho}\left(\varsigma + \frac{\mu}{3}\right)\nabla(\nabla.v) \tag{8.12}$$

with μ and ς representing the coefficients of shear and bulk viscosity as defined earlier. Along with these, the system should also satisfy the set of free field Maxwell's equations in the reduced form

$$\nabla \cdot \vec{D} = 0, \quad \nabla \cdot \vec{B} = 0, \quad \nabla \times \vec{E} = -\frac{\partial \vec{B}}{\partial t}; \quad \nabla \times \vec{H} = \vec{j}, \tag{8.13}$$

and the Ohm's law $j = \sigma\left[\bar{E} + v \times \bar{B}\right]$, σ being the conductivity. As any fluid system requires an equation of state, it can be either of these three: (a) ρ a constant, $\Rightarrow \nabla \cdot v = 0$, (b) isothermal, pV a constant or (c) adiabatic, pV^γ is constant. If the electric field is absent, then the equation (8.12), may be written as

$$\frac{dv}{dt} = F - \frac{1}{\rho}\left(\nabla p - \frac{1}{c}(\nabla \times B) \times B\right) + \frac{\mu}{\rho}\nabla^2 v + \frac{1}{\rho}\left(\varsigma + \frac{\mu}{3}\right)\nabla(\nabla.v). \tag{8.14}$$

Using the vector identity for the triple vector product, the magnetic part can be re-written as $(\nabla \times B) \times B = (B \cdot \nabla)B - \nabla(B^2/2)$, which renders the equation in the form

$$\rho\frac{dv_i}{dt} = \rho F_i - \frac{\partial}{\partial x^j}\left[\left(p + \frac{B^2}{8\pi}\right)\delta_{ij} - \frac{B_i B_j}{4\pi}\right] + \mu\nabla^2 v + \frac{1}{\rho}\left(\varsigma + \frac{\mu}{3}\right)\nabla(\nabla.v) \tag{8.15}$$

which when rewritten fully in tensorial form is given by

$$\rho\frac{dv_i}{dt} = \rho F_i - \frac{\partial}{\partial x^j}\left[\left(p + \frac{B^2}{8\pi}\right)\delta_{ij} - \frac{B_i B_j}{4\pi} + \pi_{ij}\right], \tag{8.16}$$

where π_{ij} is defined as $\pi_{ij} = \varsigma\theta\delta_{ij} + \mu\left(e_{ij} - \frac{2}{3}\theta\delta_{ij}\right)$. The equation can also be written in a compact form

$$\rho\frac{dv_i}{dt} = \rho F_i - \frac{\partial}{\partial x^j}[P_{ij} + M_{ij}], \tag{8.17}$$

where, $M_{ij} = \frac{B^2}{8\pi}\delta_{ij} - \frac{B_i B_j}{4\pi}$ and $P_{ij} = p\delta_{ij} + \pi_{ij}$.

The diagonal part of the magnetic tensor M_{ij}, which is $B^2/8\pi$, that couples to the fluid pressure is generally known as the magnetic pressure, while the non-diagonal part is sometimes referred to as tension along the field lines. (For a simple but detailed discussion on this refer to (Rai Choudhuri, 1998).

For the evolution of the magnetic field, one finds from Ohm's law and Maxwell's equations,

$$\frac{\partial B}{\partial t} = \nabla \times v \times B + \frac{1}{4\pi\sigma}\nabla^2 B. \tag{8.18}$$

In the absence of viscosity, one can, in principle, solve for the variables, v, p, B and ρ, using the above set of equations. The factor $1/4\pi\sigma$ is referred to as 'magnetic diffusivity λ, and obviously for fully conducting fluids it tends to zero as $\sigma \to \infty$.

In the context of accretion disks, it is known that for stability the angular momentum transport is effected through the Rayleigh criterion $\frac{d}{dr}(\Omega(R)^2) > 0$, when the magnetic field is absent. As Camenzind (2008) discusses, if there is a magnetic field then even a weak one destabilises the disks, for configuration with $d\Omega^2/dlnR < 0$, which seems to exist always in disk configurations. This criterion is called *magnetorotational instability*. Though it always exists the growth rate depends upon the magnetic field configuration. The existence of such an instability tells us that the ultimate model for accretion disks is magnetohydrodynamical (MHD) including some form of radiation transport. As summarized by (Camenzind, 2008) the physical reason for this instability seems to be as follows. Considering the effect of perturbing a weak vertical field threading an otherwise uniform disk, if the field remains frozen into the plasma, field lines connecting adjacent annuli in the disk will get sheared by the differential rotation into a trailing spiral pattern. If the field is weak enough, magnetic tension will not be able to bring the field lines back to the vertical and the magnetic tension acts to reduce the angular momentum of the inner fluid element and boost the outer one, providing angular momentum transport in the outward direction that is required to drive the accretion process.

Chapter 9

Plasma and Its Role in Accretion Physics

It is well known that the various phases of matter in the Universe exist at different temperatures, and with increasing temperature solids turn into liquid and on further supply of heat take the gaseous state. Further increase of temperature (at and beyond the ionisation temperature) will go to the state where the atoms get ionised thus reaching the plasma state which is a collection of ions and electrons. As accretion is a process where the matter is attracted by the star from its surroundings, this incoming matter could be normally in a gaseous state or plasma state or a mixture of both depending upon the ambient temperature.

Almost all discussions of bulk matter deal with the behaviour of matter under the influence of fields and forces prominently electromagnetic or gravitational and many times both together. Under this influence particles of matter move around randomly colliding with each other and the distance a particle travels before encountering a collision is called the mean free path usually denoted by the letter λ. Again for any system, there exists a length scale L and if L << λ, one can consider the matter under consideration to be a fluid (liquid or gas) having a definite velocity v, temperature T and density ρ defined at each point of the matter distribution, which can then be studied using fluid mechanical equations. On the other hand, if L > λ, then one has to look into the interaction between particles, and more so when it is plasma which are charged particles (ions and electrons) that could also interact with electromagnetic fields. However, as both the electromagnetic and gravitational forces are of order $1/r^2$ it is difficult to define a characteristic length scale that can be applied to all distributions. As pointed out in Rai Choudhuri (1998) in a system of charged particles at any given point apart from the locally produced electric and magnetic fields if there exists an external field one needs to solve the equations of motion of all particles simultaneously using kinetic theory which could be very difficult. For certain distributions, statistical mechanics could be useful to show that plasma in thermodynamical equilibrium follows a Maxwellian distribution, whereas fluid mechanical treatment would hold for neutral particles. If there are large magnetic fields present as it happens in astrophysical situations, when the length scale of the plasma is much greater than the gyroradius of the particles in the magnetic field, hydrodynamical treatment could work which is referred to as magneto-hydrodynamics.

One of the important features of a plasma is its charge neutrality as even a small difference in the number densities of ions and electrons would produce a large electric field which induces the movement of the charged particles to retain the charge neutrality of the plasma. If n_e is the number density of electrons which should be same as ion density, the mean distance between particles would be of the order $n_e^{-1/3}$, giving the Coulomb potential to be $e^2 n_e^{-1/3}$. If the mean kinetic energy is of the order of plasma temperature T, satisfying the condition, $n_e e^6 / T^3 \ll 1$, the associated plasma is considered as an ideal plasma. However, if there exists a small variation in the number densities resulting in charge imbalance $(n_e - n_i = n_1)$ the electrons gain velocity due to electric field which will induce motion of plasma particles, mainly that of electrons (as they are much lighter than ions) as a fluid. If v_e represents the electron fluid velocity, then one can write the Euler equation for the force balance, using the definitions

$$\rho_m = m_0 n_e, \text{ and } \rho F = -en_e E$$

and for a cold plasma ($p = 0$) the force balance yields

$$m_0 \, dv_e/dt = -e \, E. \tag{9.1}$$

The equation of continuity, which is the same as the equation for charge conservation, is then given by

$$\frac{\partial n_e}{\partial t} + \nabla \cdot (n_e v_e) = 0. \tag{9.2}$$

As $n_e = n_0 + n_1$, and $n_1 v_e$ is negligible, one gets

$$\frac{\partial n_1}{\partial t} + n_0 \nabla \cdot v_e = 0. \tag{9.3}$$

Similarly, the force equation yields

$$m_0 \frac{\partial v_e}{\partial t} = -eE. \tag{9.4}$$

Eliminating, v_e between these equations, one finally can get

$$\frac{\partial^2 n_1}{\partial t^2} + \left(\frac{4\pi n_0 e^2}{m_0} \right) n_1 = 0 \tag{9.5}$$

showing that n_1, the charge difference between electrons and ions, oscillates with a frequency

$$\omega_p = \sqrt{\frac{4\pi n_0 e^2}{m_e}} \tag{9.6}$$

that is called the *plasma frequency*. This plays a very important role while discussing the passage of electromagnetic waves through a plasma, as only the radiation with frequency $v \geq v_p = \omega_p/2\pi$ can pass through the plasma. Thus, the electron density plays a very important role in the study of astrophysical and space plasma.

Having seen the direct effect of charge imbalance, one can consider its other effect that occurs due to the grouping of charges around the opposite charge, as a result of normal coulomb interactions. This results in what is called a 'shielding', even in an unperturbed plasma due to the thermal motion of the electrons and the distance of its influence is known as Debye length or Debye radius. In order to derive an expression for this

length, one can consider the plasma to be in thermal equilibrium and satisfying Boltzmann equation, such that if φ is the electrostatic potential, the
number densities of ions and electrons are given by

$$n_i = n_0 \exp\left(\frac{-e\phi}{k_B T}\right); \quad n_e = n_0 \exp\left(\frac{e\phi}{k_B T}\right), \tag{9.7}$$

where T is the temperature and k_B is the Boltzmann constant. Substituting
in the Poisson equation, one finds

$$\nabla^2 \varphi = -4\pi \, (n_i - n_e) = e \, \varphi/\lambda_d^2, \tag{9.8}$$

where $\lambda_D = (kT/8\pi n e^2)^{1/2}$ is called the Debye length or Debye radius. The
solution of the Poisson equation is given by $\phi = (Q/r) \exp(-r/\lambda_D)$, which
indicates that the effect of the charge is screened beyond the distance λ_D.
These two scales, of time (v_p^{-1}) and of length $\lambda_D = (k_B T/8\pi n e^2)^{1/2}$ are associated to give $v_e = \lambda_D v_p$ the thermal velocity of the electrons. One thus finds
that all physical characteristics of plasmas are related to the main physical
features, the electron density n_e and the temperature T.

Figure 9.1 shows how these two vary over different types of plasma
one encounters in general. The collective effects of a plasma depend upon
the influence of the charge on the particles interacting and are of the order
$n\lambda_D^3$, which can interact simultaneously and the inverse of this number is
called the plasma parameter and is denoted by

$$g = (n\lambda_D^3)^{-1} = (8\pi)^{3/2} e^3 n^{1/2}/(k_B T)^{1/2} \tag{9.9}$$

showing that for smaller values of g the collective interaction is larger.
Before moving on, it is necessary to introduce one more concept that
could play an important role in the context of single particle dynamics —
the guiding center approximation.

Consider a charged particle in an electromagnetic field, which under-
goes acceleration. According to Newton's second law, the equations of
motion are given by, $mdv/dt = e(E + v \times B)$, where v is the velocity, E and
B being electric and magnetic fields, and the right-hand side represents the
Lorentz force. It is useful to analyse the motion in two stages. First, one
can consider the electric field to be negligible, and concentrate only on the

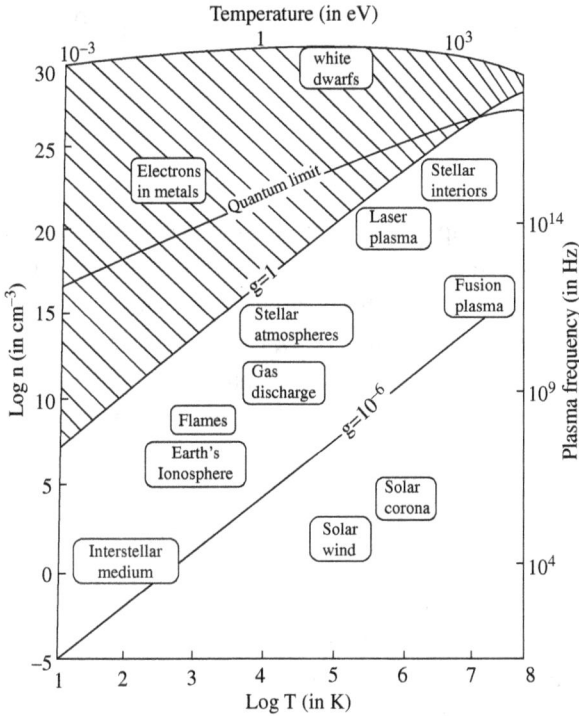

Fig. 9.1.   Adopted from Rai Choudhuri (1998).

magnetic field B. As the force is in the perpendicular direction, it does no work on the particle, conserving its kinetic energy as shown by the equation

$$v \cdot m \frac{dv}{dt} \equiv \frac{d}{dt} \left(\frac{mv^2}{2} \right) = ev \cdot (v \times B) = 0. \tag{9.10}$$

Considering the components of the velocity along and perpendicular to B, this equation may be written as

$$dv_{\parallel}/dt = 0; \quad dv_{\perp}/dt = (e/m)(v_{\perp} \times B). \tag{9.11}$$

Resolving along a Cartesian frame of reference with the magnetic field along the z-axis, one gets

$$m\dot{v}_x = eBv_y, \qquad m\dot{v}_y = -eBv_x, \qquad v_z = 0. \tag{9.12}$$

Taking time derivatives and using the same equations one finds

$$\ddot{v}_x = -\omega v_y; \qquad \ddot{v}_y = -\omega v_x, \qquad \omega = (eB/m), \tag{9.13}$$

which describes a simple harmonic motion with frequency ω, known as the *cyclotron frequency*. Thus, the particle executes a circular motion around the field line, called the Larmor orbit, whose radius is called the gyroradius or the Larmor radius. Now, if one considers the electric field E also, first consider the motion in the magnetic field by integrating the equations using the initial condition that the particle starts from the point (x_0, y_0) at $t = 0$, such that

$$x = x_0 + r_L \sin\omega t, \qquad y = y_0 + r_L \cos\omega t$$

with the Larmor or gyroradius $r_L = |mv_\perp/eB|$. The center of the circle (x_0, y_0) around which the particle orbits is called the *guiding center*. As there also exists a constant electric field E, in the x–z plane, the original equation will have additional terms,

$$m\dot{v}_x = eBv_y + eE_x, \qquad m\dot{v}_y = -eBv_x, \qquad v_z = eE_z. \tag{9.14}$$

Using a similar approach as given above, one can easily get the system of equations to be

$$(D^2 + \omega^2)(v_y + E_x/B) = 0; \quad (D^2 + \omega^2)(v_x) = 0; \quad (D - \omega v_z/B) = 0. \tag{9.15}$$

$D \equiv d/dt$, with solutions

$$v_x = v_\perp e^{i\omega t}, \; v_y = \pm iv_\perp e^{i\omega t} - E_x/B, \; v_z = (eE_z/m)t + E_z. \tag{9.16}$$

showing that the particle apart from executing the Larmor motion suffers a drift in the direction of the vector $E \times B$, which in the present case is along the y-axis. This shows that in space the particle traces a helical path. This path is interpreted as the particle moving around the guiding center while the guiding center has a drift along the direction of the electric field. In fact, this drift is referred to as inertial drift, which can be also caused by any other source like a *gravitational field* instead of the electric field. An important feature of this approximation is the identification of adiabatic invariants associated with particle motion in the interacting fields. As demonstrated by Schmidt (1966) the magnetic moment, $\mu_m = qv_\perp R/2$, is a constant of motion known as an *adiabatic invariant*, with R being the gyroradius. It may also be noted that the definitions of the gyro radius and of the gyro frequency given above imply that for the case when particle velocity $v_\parallel = 0$, the particle's circular path is due to the charged particle moving in a magnetic field, the time average of which gives rise to a ring current, $I = q^2B/2\pi m$, corresponding to the magnetic moment $\mu_m = I\pi R^2 = BR^2q^2/2m$. Further, if the magnetic flux surrounded by the path is ϕ then $\mu_m = (q^2/2\pi m)\,\phi$ showing that the magnetic moment is proportional to the enclosed flux and thus it will also be a constant of motion. This implies that the magnetic field is frozen into the particle path, giving rise to the concept of 'frozen in magnetic field', a characteristic exhibited by several situations in space and astrophysical plasmas. giving the magnetic field strength $B \approx \sqrt{\dot{M}}\,(GM)^{1/4}\,r^{-5/4}$.

For black holes as $r = R_g$, for critical accretion rate $\dot{M}_c = L_{edd}/c^2 \cong 1.4 \times 10^{17} M/M_{sun}$, if the black hole mass is $\approx 10 M_{sun}$, the magnetic field strength reaches almost about 5×10^7 gauss. In view of the fact that such high magnetic fields could exist in the vicinity of compact objects, it is necessary to study the dynamics of charged particles in electromagnetic fields on curved spacetime. As was in the case of pure gravitational fields, it would be easier to study the particle dynamics in terms of the effective potentials on the basic geometry with painted magnetic fields. This requires a discussion related to electromagnetic fields on curved spacetime which will be considered next.

Chapter 10

Electromagnetism on Curved Space-Time

The significance of special relativity for discussing electric and magnetic fields is self-explanatory as the development of Maxwell's equations need Lorentz invariance which is the integral consequence of special relativity. What is the relevance of general relativity while looking at the cosmic magnetic fields? In an astrophysical scenario, one is looking at the fields associated with heavy mass objects, with strong gravitational fields one is trying to discuss these fields superimposed on gravitational fields described by curved space-time of general relativity. Further, the material distribution one is dealing with is at such high potential that it could be in ionised form like a plasma (ions and electrons). To discuss the dynamics of such a configuration one needs to formulate the dynamical equations of electromagnetism on curved space-time as given by the associated differential manifold. As the electromagnetic field is a conservative field, the forces are derivable from a potential. Since we are working in a four-dimensional space-time manifold, one defines an electromagnetic four potential A_i, such that one can define an anti-symmetric four tensor $F_{ij} = A_{j;i} - A_{i;j}$, with the semicolon representing a covariant derivative, as introduced earlier. One can then choose the Lagrangian density

$$\mathcal{L} = \sqrt{-g} \ (R + \kappa (F^{ij} F_{ij} + \alpha A_i j^i), \tag{10.1}$$

with R representing the scalar gravitational curvature and F_{ij} the electromagnetic field tensor, A_i is the vector potential, $j^i = (j^a, \phi)$ the four-current vector and g the determinant of the metric tensor g_{ij}. Considering the action $J = \int \mathcal{L} d^4 x$, the variation of action with respect to the metric g, $(\delta J)_g = 0$ gives the equations, (Prasanna, 2017)

$$R_{ij} - (1/2) R\, g_{ij} + k(F_{il}F_j^l - (1/4)\, g_{ij}\, F_{kl}\, F^{kl}) = 0 \tag{10.2}$$

while the variation with respect to the potential A, $(\delta J)_A = 0$ gives

$$F_{;j}^{ij} = j^i. \tag{10.3}$$

As it was in the case of special relativity, the anti-symmetry of the field tensor F_{ij} yields the cyclic identity $F_{(ij;k)} = 0$, on the Riemannian background where the metric connection is symmetric $\left(\Gamma_{ij}^k = \Gamma_{ji}^k\right)$. Thus one gets the same relation as in special relativity.

$$F_{ij,k} + F_{ki,j} + F_{jk,i} = 0. \tag{10.4}$$

On the other hand on a general differential manifold where the connection can be asymmetric, the relation gives with anti-symmetric part, $\left(\Gamma_{ij}^k = -\Gamma_{ji}^k\right)$, as torsion is non-zero the equation

$$F_{ij,k} + F_{ki,j} + F_{jk,i} + Q_{ik}^l F_{jl} + Q_j^l F_{il} + Q_{ij}^l F_{kl} = 0 \tag{10.5}$$

Q_{ij}^k being the Contortion tensor defined by $Q_{ij}^k = \left(\Gamma_{ij}^k = -\Gamma_{ji}^k\right)$

10.1 Single Particle Dynamics

Even with the magnetic fields, an uncharged particle still moves on a geodesic whose dynamics may be discussed in terms of the geodesics of Schwarzschild geometry as follows.

Considering the simplest of the solutions of Einstein's equations, the Schwarzschild geometry, one has the space-time metric as given by

$$ds^2 = \left(1 - \frac{2m}{r}\right) dt^2 - \left(1 - \frac{2m}{r}\right)^{-1} dr^2 - r^2 d\theta^2 - r^2 \sin^2\theta d\varphi^2 \tag{10.6}$$

which represents the external geometry of a static, spherically symmetric compact star. It can be easily seen that for any particle moving in this field because of the static and spherically symmetric nature of space-time, the energy E and the angular momentum l are constants of motion expressed as given by

$$c^2\left(1-\frac{2m}{r}\right)\dot{t} = E, \quad r^2\dot{\varphi} = l. \tag{10.7}$$

As the metric itself provides one of the first integrals of motion, it can be verified that for the motion confined to the equatorial plane ($\theta = \pi/2$), one has on using the above constants, in the metric, the relation

$$(dr/ds)^2 = E^2 - \left(1-\frac{2m}{r}\right)\left(1+\frac{l^2}{r^2}\right). \tag{10.8}$$

The 'effective potential, defined as the energy of the particle at the turning points ($dr/ds = 0$) is of the form

$$V_{eff}^2(r) = \left(1-\frac{2m}{r}\right)\left(1+\frac{l^2}{r^2}\right). \tag{10.9}$$

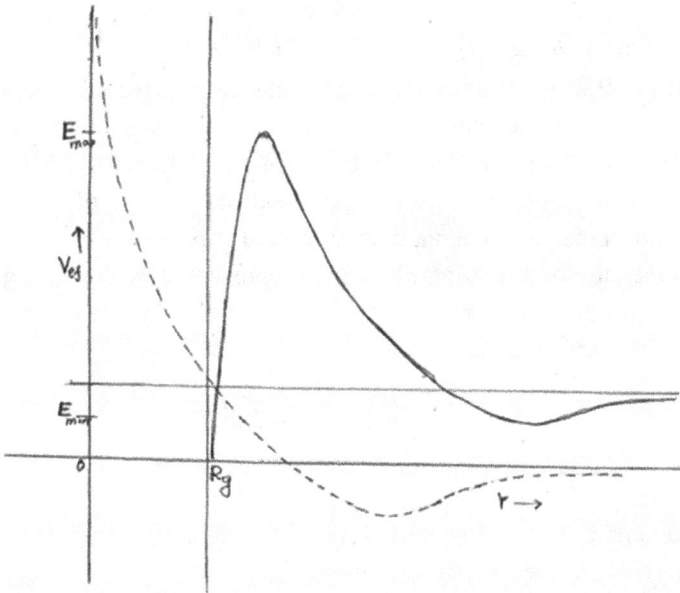

Fig. 10.1.

Figure 10.1 depicts the plots of V_{eff} as a function of the coordinate r both for the Newtonian (dashed curve) V_{eff}, and the general relatvistic (solid curve) cases.

As can be seen, the centrifugal barrier increases steadily in the Newtonian case whereas in general relativity it increases to a maximum near the compact object and then decreases to zero at the event horizon ($r \rightarrow 2m$) of the blackhole. This means while energy increase cannot support accretion (particle streaming in) as viewed from the Newtonian perspective, in GR particles with energy higher than maximum of V_{eff} can cross the barrier over and accrete onto the central star. Only those particles with energy E, $1 < E < V_{max}$ will have hyperbolic orbits as they are deflected by the centrifugal barrier. An additional feature present here but not in the Newtonian dynamics is the fact that particles can have unstable circular orbits as shown by the maximum of V_{eff}. From the ensuing equation, one can see that $dV/dr = 0$, implies $mr^2 - (r - 3m)l^2 = 0$, which in turn gives $r = (l^2/2m)\left(1 \pm \sqrt{\left(1 - \frac{12m^2}{l^2}\right)}\right)$ that corresponds to the locations of maximum and minimum of V_{eff} for $l^2 \neq 12m^2$.

For $l = 2\sqrt{3}\ m$, $r = 6m$ corresponds to the last stable orbit for a particle and for $l < 2\sqrt{3}\ m$, there exist no orbits. The energy at the last stable orbit is $E = (r - 2m)/\sqrt{(r(r - 3m))}$. This gives E at $r = 6m$ to be $\approx 0.943\ m_0c^2$, yielding the binding energy of the particle at the last stable orbit to be $E_{bind} \approx 1 - 0.943 = 0.057 m_0c^2$. Thus in the Schwarzschild geometry (for static black hole) the energy released on accretion is about 5.7% which is much greater than what one finds in spherical accretion. However, the situation changes to a better prospect both in the case of rotating black holes and in the case when a magnetic field is present.

In the case of a rotating black hole represented by the Kerr geometry, the effective potential for a particle in the equatorial plane can be obtained through the first integrals,

$$u^t = [r^3 + a^2r + 2ma^2)E - 2aml]/r\Delta \tag{10.10}$$

$$\text{and } u^\varphi = [(r - 2m)l + 2amE]/r\Delta. \tag{10.11}$$

For a particle with rest mass zero on the equatorial plane ($\theta = \pi/2$), one can see that the metric can be expressed as

$$R(E,l,r) \equiv \dot{r}^2 r^2 = E^2(r^3 + (r + 2m)a^2) - 4am\ EL - (r - 2m)l^2 - m_0^2 r\Delta.$$

The effective potential given by the energy at the turning point $\dot{r} = 0$, may be factorised as $(E - V_+)(E - V_-)$ with V given as

$$V_{\pm} = \frac{[-2mal \pm \Delta^{\frac{1}{2}}\,[r^2 l^2 + m_0^2 r\rho]^{1/2}]}{\rho}, \quad \rho = r^3 + a^2(r + 2m). \quad (10.12)$$

Choosing the + sign to match with the Schwarzschild case, when $a = 0$, one gets on the event horizon, where $\Delta = 0, V_+ = -\frac{2aml}{\rho}$. For a slowly rotating black hole with terms of order a^2 and higher being neglected, $V_+ = V_{sch} + 2aml/r^3$. Further on the horizon, V_{eff} is negative for a particle with direct orbit, meaning the angular momentum l and a are of the same sign and it is positive for particles in a retrograde orbit (a and l are of opposite sign). As pointed out by Bardeen (1972), for a given value of the angular momentum at a given radius the physically accessible region ($R(r) \geq 0$) is limited to only those values of energy $\geq E_{min} = V_+$. A typical plot of the effective potential for a given l and a is as shown in Fig. 10.2.

For particles in circular orbits with ($\frac{\partial R}{\partial r} = 0$, and $R = 0$). The energy and angular momentum are given by Bardeen *et al.* (1972).

$$E = \frac{(r^{\frac{3}{2}} - 2mr^{\frac{1}{2}} \pm a\sqrt{m})}{A}, \quad l = \pm\frac{\sqrt{m\left(r^2 \mp 2a\sqrt{mr} + a^2\right)}}{A}, \quad \text{and} \quad (10.13)$$

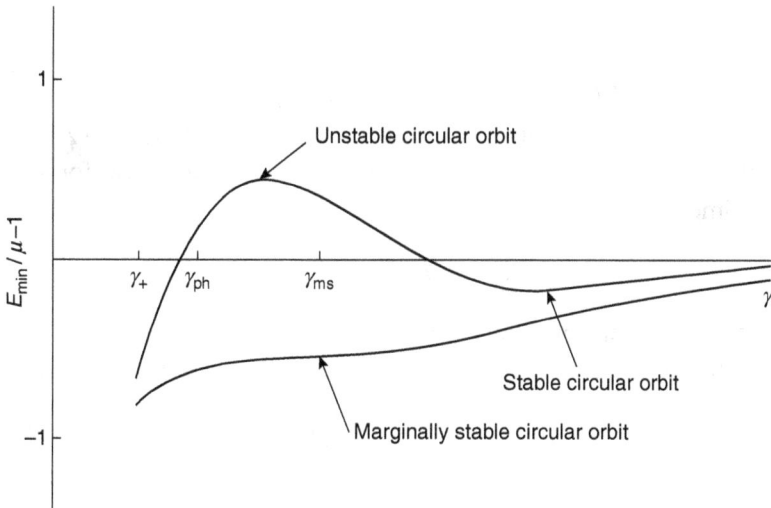

Fig. 10.2. Schematic diagram of the effective potential for Kerr metric (Bardeen 1972).

$$A = r^{3/4} \sqrt{(r^{3/2} - 3mr^{1/2} \pm 2a\sqrt{m})}.$$

In the above, the two signs for l distinguish the cases $al > 0$ (co-rotating), and $al < 0$ (counter-rotating). We now turn our attention to the case of charged particle orbits with electromagnetic fields on a curved background as they play a more important role in accretion dynamics as the accreting matter would mostly be, ions and electrons.

10.2 Charged Particle Dynamics on Curved Space Time

As the most general black hole space time is represented by the axi symmetric, stationary Kerr-geometry there exist two Killing vectors the time-like one ξ^i and the space-like one ζ^i such that in the absence of any other field other than gravity for any particle there are two constants of motion as given by the energy $E = -\xi^i u_i$ and the angular momentum $l = \zeta^i u_i$, u^i being the four velocity of the particle. E and l are constants along the geodesics $u^i_{;j} u^j = 0$. If there exists an electromagnetic field respecting the same symmetry as the background geometry characterised by the vector potential A_i then for a charged particle with charge e in that field the generalised first integrals are given by the relations

$$(u_i + eA_i)\xi^i = -E \text{ and } (u_i + eA_i)\zeta^i = l,$$

where now E and l are constants along the trajectories $u^i_{;j} u^j = eF^i_j u^j$, $F_{ij} = A_{j;i} - A_{i;j}$, if and only if A_i satisfies the Lie transport equation $A_{i;j} K^j + A_j K^j_{;i} = 0$, for all Killing vectors of the space time. The general metric for such a space-time may be written as

$$ds^2 = g_{tt} dt^2 + 2g_{t\varphi} dt\, d\varphi + g_{rr} dr^2 + g_{\theta\theta} d\theta^2 + g_{\varphi\varphi} d\varphi^2$$

g_{ij} s being functions of r and θ only.

With this, the equations for the constants of motion take the form

$$g_{tt} u^t + g_{t\varphi} u^\varphi = -(E + eA_t), \tag{10.14}$$

$$g_{\varphi t} u^t + g_{\varphi\varphi} u^\varphi = (l - eA_\varphi). \tag{10.15}$$

Solving these one finds the first integrals to be

$$u^t = [g_{\varphi\varphi}(E + eA_t) + g_{t\varphi}(l - eA_{\varphi})]/D, \tag{10.16}$$

$$u^\varphi = [g_{t\varphi}(E + eA_t) + g_{tt}(l - eA_{\varphi})]/D, \tag{10.17}$$

where $D = g_{\varphi\varphi}g_{tt} - g_{t\varphi}^2$.

As u^i satisfies the condition of ortho normality $g_{ij}u^i u^j = \pm 1$, using the above expressions and solving for u^r one gets,

$$(u^r)^2 = -(g_{rr}D)^{-1}[D + g_{tt}(l - eA_{\varphi})^2 + g_{\varphi\varphi}(E + eA_t)^2$$
$$+ 2g_{t\varphi}(E + eA_t)(1 - eA_{\varphi})]. \tag{10.18}$$

Since the effective potential is defined as the energy of the particle at the turning point where $u^r = 0$, one gets V_{eff} to be

$$V_{\mathit{eff}} = E_\pm = -\left[eA_t + (g_{t\varphi}/g_{\varphi\varphi})(l - eA_{\varphi}) \right] \pm g_{\varphi\varphi}^{-1}\sqrt{-\Delta}. \tag{10.19}$$

with $\Delta = D\{g_{\varphi\varphi} + (l - eA_{\varphi})^2\}$.

The $\pm$ sign indicates the positive and negative energy states for the particle with different domains of dependence. While analysing the orbits one in principle considers the solutions for both the metric potential g_{ij} and the vector potential A_i, of the set of Einstein-Maxwell equations,

$$R_{ij} - \frac{1}{2}Rg_{ij} = -\chi E_{ij}; \quad E_{ij} = F_{ik}F_j^k - \frac{1}{4}g_{ij}F^{kl}F_{kl}, \tag{10.20}$$

with F_{ij} satisfying the Maxwell's equations on curved space

$$F_{(ij,k)} = 0 \text{ and } F_{;k}^{ik} = J^i. \tag{10.21}$$

However, these equations being a combined set of coupled nonlinear second-order partial differential equations are very difficult to solve analytically. Two exact solutions are known as given by the Kerr-Newman solution and the Reissner–Nordstrom solution one representing a charged rotating black hole and the other charged static black hole. Since we are interested in solutions with magnetic fields one needs to look for situations where there could be an overlapping magnetic field which may not contribute to the geometry. Considering the case of neutron stars which are endowed with high magnetic fields one can see that the magnetic field energy even for a 10^{12}G field is far less as compared to the gravitational potential energy of even a one solar mass star. Thus, while considering Einstein's equations, one can ignore the contribution from the electromagnetic field energy source term ($E_{ij} = 0$) and consider only the equations $R_{ij} = 0$. For example, in the vicinity of a star the geometry can be taken as that of Schwarzscild solution for non-rotating and that of Kerr for rotating stars. However, for solving the corresponding Maxwell's equations on the curved background $\left(\sqrt{-g}\, F_i^{\,j}\right)_{,j} = 0$, one uses the metric potentials g_{ij} of the Schwarzschild or the Kerr metric as the need be. Basing on these arguments one can now write the expression for the four potentials for an asymptotically dipolar magnetic field superposed on curved background to be $A_i = (0, 0, A_\varphi, A_t)$ with

$$A_\varphi = \frac{3\mu\sin^2\theta}{8m^2}\left[r^2\ln\left(1-\frac{2m}{r}\right)+2m(r+m)\right], \quad A_t = 0. \qquad (10.22)$$

for the Scwarzschild case (non-rotating blackhole) (Prasanna and Varma, 1977) while for the case of rotating blackholes with Kerr space-time the components of the four potential are (Prasanna and Vishveswara, 1978; Peterson, 1975).

$$A_\varphi = \left(\frac{-3\mu\sin^2\theta}{4\gamma^2\Sigma}\right)\{r(r-m)a^2\cos^2\theta+r(r^2+mr+2a^2)-[r(r^2+a^2)$$

$$(r-2m)\Delta a^2\cos^2\theta](1/2\gamma)\ln(r-m+\gamma/r-m-\gamma)\}, \qquad (10.23)$$

and

$$A_t = \left(\frac{-3\mu a}{2\gamma^2\Sigma}\right)\{-r(r-m\cos^2\theta)+[r(r-m)+(a^2-mr)\cos^2\theta]$$

$$(1/2\gamma)\ln(r-m+\gamma/r-m-\gamma)\};\gamma^2 = m^2 - a^2.$$

One can see that in the rotating case along with the dipolar magnetic field a quadrupolar electric field also gets generated. Using these expressions for the four potentials, one can write the effective potential for the case of a charged particle in the field of an electromagnetic field surrounding a rotating blackhole as given by

$$V_{eff} = -A_t + K/R,$$

with $K = 2\alpha(L - \overline{A_\varphi}) \pm \Delta^{1/2}\{\rho^2(L - \overline{A_\varphi})^2 + \rho R\},$ wherein

$$R = \rho^2 + \alpha^2(\rho + 2), \quad \Delta = \rho^2 + \alpha^2 - 2\rho, \text{ and} \qquad (10.24)$$

$$\rho = r/m, \alpha = a/m, \overline{A_\varphi} = A_\varphi/m.$$

For more details in this context, one may refer to Prasanna (1980, 2015).

As an example, we consider the case of a dipole field super imposed on the Kerr space-time as given by the four-potential (10.23) and the corresponding effective potentials V_{eff} (10.24). represented by the adjoining figure (10.3) depicting the potential V_+ as a function of r in the equatorial plane for different values of the parameters a, λ, and L. It may be noticed that the potential has four extrema. (Due to scaling difficulties the inner maximum and the outer most minimum are not shown) (Prasanna and Vishveswara, 1978).

As λ increases the value of the inner maximum (M1) increases while that of the inner minimum (m1) and outer maximum (M2) decrease for fixed a and L monotonically. This amounts to the fact that the potential well created by the maxima decreases in depth thus reducing the possibility of trapping of particles with lower energy. On the other hand for fixed a and λ as L increases the centrifugal barrier gets larger as M2 increases.

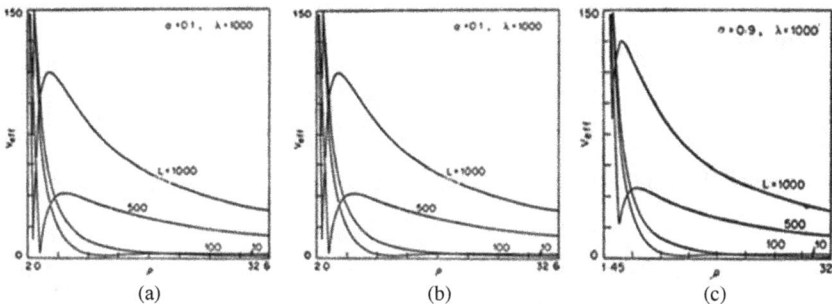

(a)   (b)   (c)

Fig. 10.3.

Due to all these different features for fixed values of the parameters, the potential well appears either fully inside or outside the ergo surface which has a significance as pointed out by (Prasanna and Vishveswara, 1978). As the particles trapped in a potential well have closed orbits, the gyrating orbits exhibit different features for the following reasons. Inside the ergo-sphere, the particles cannot gyrate due to the effect of frame dragging by the compact object. As illustrated in Fig (10.4), the projection of orbits on the equatorial plane illustrates this feature.

Particle orbits in the equatorial plane (Kerr with dipole), showing the varying degrees of gyration, depending upon the magnetic field (λ) and the angular momentum L. ρ_0, ρ_1, ρ_2, indicate the initial position and the two turning points.

In order to justify this claim Prasanna and Chakraborty (1980) performed the required analysis in the LNRF (locally non-rotating frame, also known as ZAMO. Figure 10.5 shows for the same parameters the projection of orbits for ZAMO as depicted in Fig (10.4) in B–L coordinates.

Comparing with orbits in Figure (10.4), one can see, for the same parameters while there was no gyration depicted in the earlier case, here the gyrations are seen. (Prasanna, 2017). A simple explanation for this feature is as follows. When a particle gyrates around a magnetic field, in the presence of a central rotating star, the particle will be executing part of the time prograde motion with respect to the central star and the other times retrograde. As retrograde motion is not possible within the ergosur-face the particle is dragged along without gyration. As argued in (Prasanna and Vishveswara, 1978), if the particle has to gyrate then its angular velocity $d\varphi/ds$ has to be zero at some point say r_1. From (10.11) one should then have,

$$(r_1 - 2)(L - A_\varphi) + 2a = 0. \tag{10.25}$$

Using this in the expression for the radial velocity dr/ds, one finds

$$(dr/ds)^2 = \left(\frac{\Delta}{r_1^2}\right)\left[(1 - 2m/r_1)^{-1}\{(E + A_t)^2 - 1\}\right]. \tag{10.26}$$

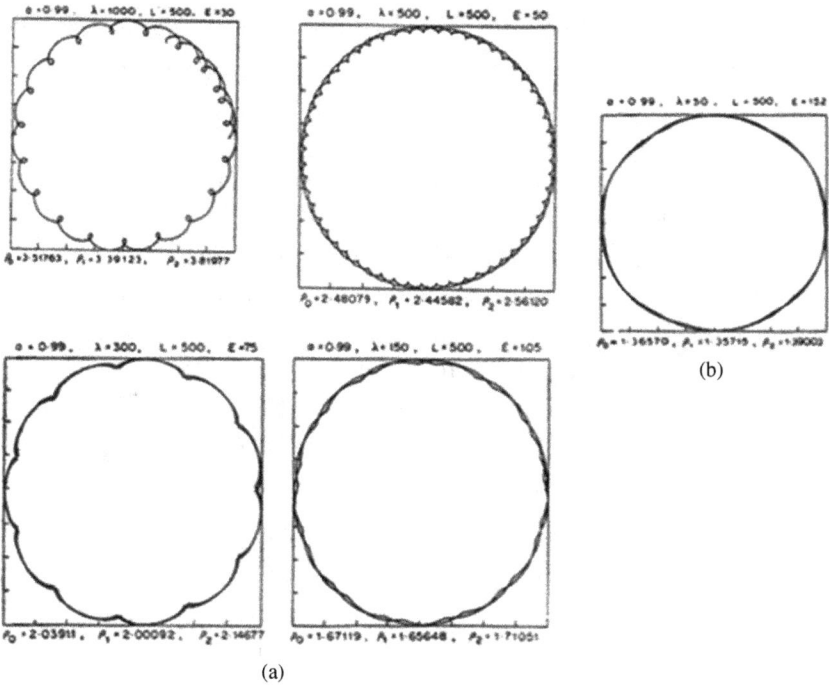

Fig. 10.4. Adopted from Prasanna and Vishveswara, (1978).

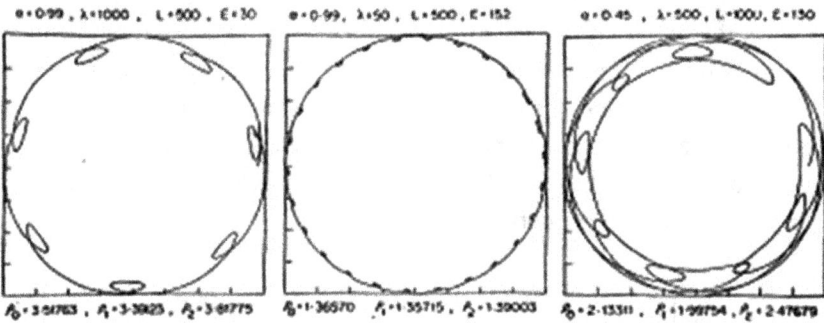

Fig. 10.5. Particle orbits in the equatorial plane of Kerr spacetime with dipole field, as viewed from a locally non-rotating frame, also called ZAMO. Adopted from Prasanna and Chakrbaty (1980).

As Δ is positive outside the horizon, for the radial velocity to be real one should have $\left(1 - \frac{2m}{r_1}\right)^{-1} (E + A_t)^2 > 1$ which can be true only for $r_1 > 2m$, outside the ergo surface.

As most of the astrophysical bodies (except perhaps the millisecond Pulsars) are slowly rotating it is useful to consider the case of particle dynamics for slowly rotating compact objects.

Hartle and Thorne (1969) have considered such a scenario where they obtain to a certain approximation the external metric for the gravitational field of a slowly rotating, and centrifugally deformed star precisely up to second order in angular momentum a but to first order in quadrupole moment Q_d in the limit J^2/r^4, MJ/r^3, MQ_d/r^3 all being negligible (tending to zero)

$$
\left[\left(1 - \frac{2M}{r} + \frac{2Q_d}{r^3} P_2(\cos\theta)\right)dt^2 - \left[\left(1 - \frac{2M}{r} + \frac{2Q_d}{r^3} P_2(\cos\theta)\right]^{-1} dr^2 - \right.
$$

$$
\left(1 - \frac{2Q_d}{r^3} P_2(\cos\theta)\right]r^2 \left\{d\theta^2 + \sin^2\theta\left(d\varphi - \frac{2J}{r^3}dt\right)^2\right\}. \tag{10.27}
$$

As pointed out by Zeldovich and Novikov (1971) this metric has no singularity either on the Schwarzschild surface ($g_{tt} = 0$) or on the horizon ($g_{rr} = 0$), if and only if the quadrupole moment Q_d, the angular momentum J and the star's mass M satisfy the algebraic identity $Q_d = J^2/M$ (Thorne, 1971). For discussing the particle dynamics in a painted magnetic field on such a geometry, Prasanna and Gupta (1997) have considered a simpler version of this metric by neglecting the higher order terms in J and Q_d also, that results in the metric

$$
ds^2 = \left(1 - \frac{2m}{r}\right)dt^2 - \left(1 - \frac{2m}{r}\right)^{-1} dr^2 - r^2[(d\theta)^2 - \sin^2\theta(d\varphi - \omega dt)^2] \tag{10.28}
$$

which is the modified Schwarzschild metric that includes the effect of 'dragging of inertial frames' as obtained earlier by Lense and Thirring (1917). Solving the Maxwell's equations on this geometry (Prasanna and Gupta, 1997) have obtained the solution for magnetic and electric field components to be

$$B_r = \left(-3\mu\cos\vartheta/4m^3\right)\left\{\ln\left(1-\frac{2m}{r}\right)+\frac{2m}{r}+\frac{2m^2}{r^2}\right\},$$

$$B_\theta = \left(\frac{3\mu\sin\theta}{4m^2 r}\right)\left\{(1-2m/r)^{-1}+\frac{r}{m}\ln(1-2m/r)+1\right\}(1-2m/r)^{1/2} \quad (10.29)$$

$$E_r = \left(-\frac{2\mu\omega}{3r^2}\right)\{P_2(\cos\theta)(1+3m/r)\} \quad (10.30)$$

$$E_\theta = \left(-\frac{\mu\omega}{r^2}\right)\{\sin 2\theta)(1+3m/2r)\}$$

$$J^t = \left(\frac{2\mu\omega}{r^3}\right)\{P_2(\cos\theta)(1+3m/r)\} \quad (10.31)$$

With these components, one can derive the effective potential on the equatorial plane $\theta = \frac{\pi}{2}$, to be

$$V_{eff} = \frac{J\lambda}{r^4}\left(\frac{1}{3}+\frac{4}{5r}\right)+\frac{2JA}{r^2}\pm\left[\left(1-\frac{2m}{r}\right)(1+A^2)\right]^{1/2} \quad (10.32)$$

with $A = \{L/r +(3\lambda r/8)[\ln(1-2m/r) +2m/r +2m/r^2\}$ and $\lambda = \frac{e\mu}{m^2}$.

Figure 10.6 shows some typical plots of the effective potential which do indicate trends for both bound and unbound orbits. As was found in the case of Kerr background with a dipole magnetic field, there are four extrema governed by the magnetic field and the centrifugal barrier. For sufficiently high values of these two parameters, the potential well occurs quite far away from the central star, creating particle trapping and thus creates the magnetosphere for the compact object. This helps in creating the neutron star magnetosphere, a requirement for Pulsars. However, very high energy particles have only plunge orbits, particularly when L and λ are of opposite sign and $J \geq 0$. As λ depends upon both the charge e and the magnetic moment μ, it appears that for stable bound orbits, both should be of the same sign for $L < 0$. From possible details of analysis, it appears that the counter-rotating particles see deeper potential well as compared those corotating. For more details and plots of actual orbits on the equatorial plane, one can see Prasanna and Gupta (1997).

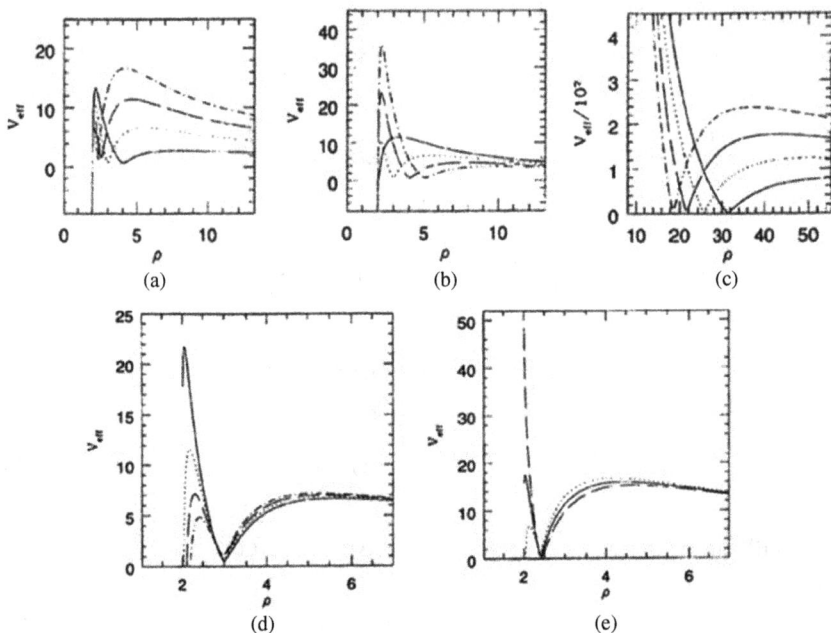

Fig. 10.6. Plots of $V_{eff}(\rho)$ for different sets of parameters. (a) $\lambda = 100$, $J = 0.310$, $L = 40$, (solid line), $= 70$ (dots), $= 100$ (dash), $= 130$ (dash-dot); (b) $L = 70$, $J = 0.310$, $\lambda = 25$, (solid line), $= 100$ (dots), $= 175$ (dash), $= 250$ (dash-dot); (c) $J = 0.310$, $\lambda = 3.10^{10}$, $L = 10^9$, $= 1.25 \times 10^9$ $= 1.5 \times 10^9$, $= 1.75 \times 10^9$; (d) $\lambda = 100$, $L = 70.78$, $J = 0$ (solid), $= 0.271$ (dots), $= 0.541$ (dash), $= 0.813$ (dash-dot); (e) Each curve represents two different sets of values, $J = 0$, $L = \pm130$, $\lambda = \pm100$, for the solid line, $J = \pm0.31$, $L = \pm130$, $\lambda = \pm100$, ($J < 0$, L and $\lambda > 0$ and vice versa) for the dashed line, and all the three positive or negative for the dotted curve (Prasanna and Gupta, 1997).

10.3 Motion of Charged Particles off the Equatorial Plane

The discussion so far concentrated on the motion of particles on the equatorial plane of the central object. In astrophysical situations it is necessary to look at the particle orbits off the equatorial plane too. Such studies were first made by Stormer in the early twentieth century, in the context of cosmic ray particles in the geomagnetic fields (but only in the Newtonian regime) and later followed by several others again in non-relativistic regimes only. As mentioned, Prasanna and Varma (1977) studied the charged particle motion in the presence of gravity using a general relativistic formulation (dipole magnetic field superposed on the Schwarzschild geometry), some details of which were discussed above. (As the orbit equations also involve transcendental functions that require numerical integration, one expresses all parameters and variables in dimensionless form using the length scale $m = MG/c^2$). They have considered the orbit equations for the four vectors $u^i = dx^i/ds$ ($i = r$, θ and φ) and the metric itself as the fourth integral, with the initial conditions, $(d\varphi/d\sigma)_0 = 0$ at $\rho = \rho_0$, and $\theta_0 = \pi/2$, that gives

$$E = \frac{-3\lambda\varrho_0^2}{8}\left[\ln\,(1-2/\rho_0)+\frac{2}{\rho_0}\left(1+\frac{1}{\rho_0}\right)\right], \tag{10.33}$$

$$(d\rho/d\sigma)_0^2 = E^2 - (1-2/\rho_0)\left\{1+\left(\rho\frac{d\theta}{d\sigma}\right)_0^2\right\} \tag{10.34}$$

$$(d\theta/d\sigma)_0^2 = \left\{\frac{E^2}{\rho_0^2}(1-2/\rho_0)^{-1} - 1/\rho_0^2\right\} \tag{10.35}$$

Specifying particular values for E, λ, ρ_0, and $(d\theta/d\sigma)_0$ one can integrate numerically the relevant equations and one can plot the trajectories. Figure (10.7) shows a typical plot for the motion off the equatorial plane.

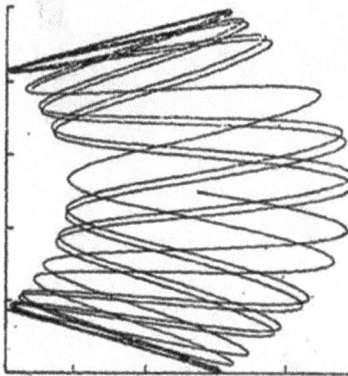

Fig. 10.7. Projection of the $r\,\theta$ motion of a positively charged particle in a dipole field on Schwarzschild background off the equatorial plane projected on the *X–Y* plane (Prasanna and Varma, 1977).

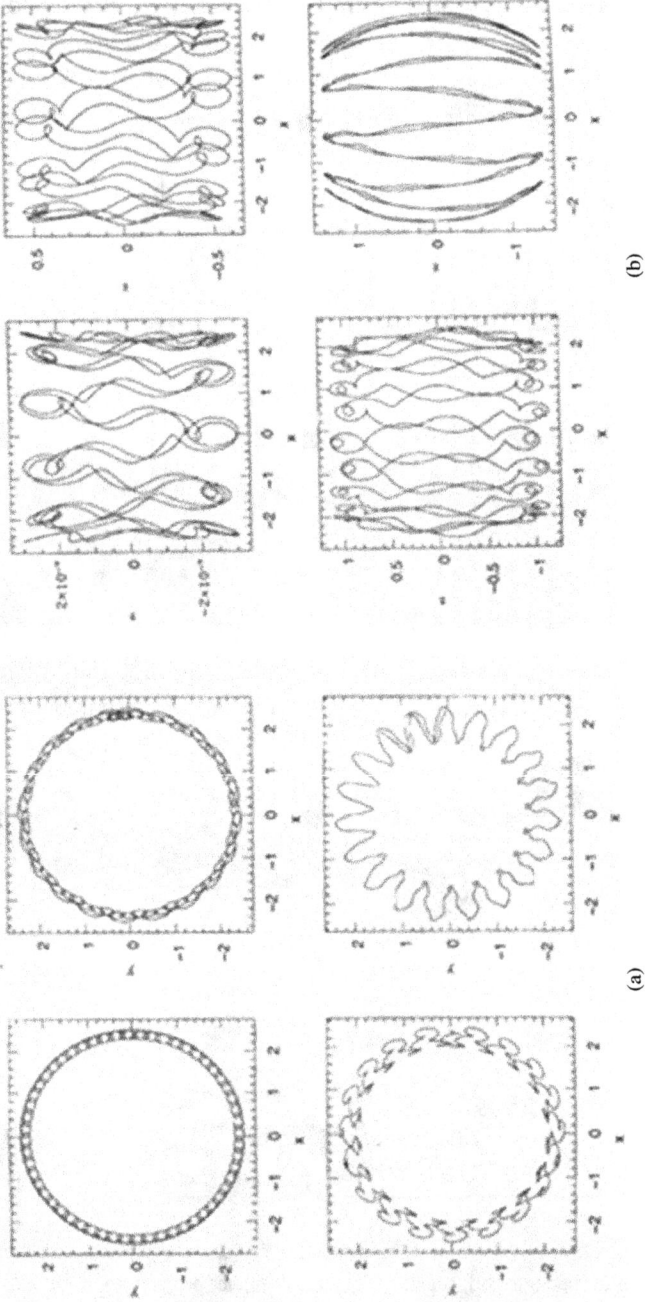

Fig. 10.8.   Projection of the particle motion, of a charged particle in a dipole field on Hartle-Thorne background off the equatorial plane, on both (X, Y) (a) and (X, Z) (b) planes. Adopted from Prasanna (1980).

As may be seen the features are almost similar to the case of a dipole magnetic field where the particle gyrates around the field line reflecting between two mirror points, on either hemispheres as to be expected when the magnetic field is large. If the field strength is not big then the particle oscillates up and down the $\theta = \frac{\pi}{2}$ plane till the value of ρ reaches a minimum when the particle gets sucked in by the central body. If one considers the case of the charged particle motion in the field of a slowly rotating star, using the Hartle-Thorne solution, with a super imposed dipole magnetic field described above, the plots are as in Fig. (10.8).

Given a small initial velocity in the θ-direction the particle executes oscillatory motion with or without gyration depending upon the location of the trapping surface with respect to the ergosurface (frame dragging surface).

10.4   Motion in a Toroidal Magnetic Field

As the discussions on the charged particle motion in magnetic fields are being taken up in the context of accretion physics, considering only dipolar field may be too restrictive and as such it is natural to look for the motion in the presence of toroidal magnetic field too. In order to consider the complete scenario, one needs to solve the basic electromagnetic field equations on a given curved background supporting gravity. In the absence of electric fields and currents, the governing equations are given by $F^{ij}_{;j} = 0$, which can be rewritten as $\partial/\partial x^j \left(\sqrt{-g}\, F^{ij} \right) = 0$. As a simple example considering the case of Schwarzschild background, one finds the equations,

$$\partial/\partial r(r^2 \sin \theta \, F^{r\theta}) = 0, \quad \partial/\partial \theta(r^2 \sin \theta \, F^{\theta r}) = 0,$$

$$\partial/\partial r(r^2 \sin \theta \, F^{r\varphi}) = 0, \quad \partial/\partial \theta(r^2 \sin \theta \, F^{\theta\varphi}) = 0,$$

and from the set $\quad F_{(ij,k)} = 0, \quad \partial/\partial \theta(F_{\varphi r}) + \partial/\partial r(F_{\varphi\theta}) = 0$

Solving these (Prasanna and Sengupta, 1994) obtained the solutions for the components as given by the toroidal component:

$$B_\varphi = B_0 \left(1 - 2m/r_0\right)/(1 - 2m/r)\sin \theta,$$

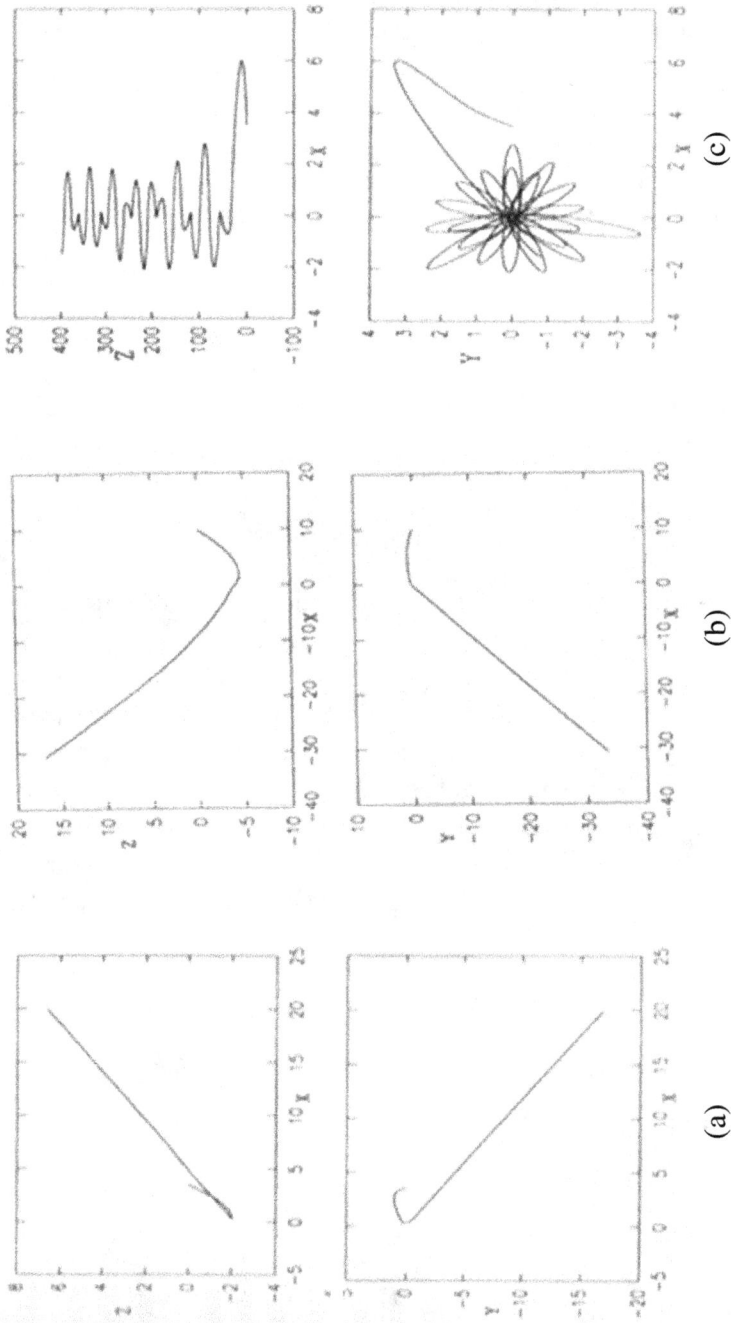

Fig. 10.9. Projections of the particle trajectories in a purely toroidal (λ_t) magnetic field, (a) E = 2, L = 10, λ_t = 50, ρ_0 = 3.5, $(d\theta/d\sigma)_0$ = 0, $(d\phi/d\sigma)_0$ = −1.92 (b) E = 5, L = 20, λ_t = 25, ρ_0 = 10, $(d\theta/d\sigma)_0$ = 0.3, $(d\phi/d\sigma)_0$ = −3.714 (c) E = 5, L = 20, λ_t = 50, ρ_0 = 3.5, $(d\theta/d\sigma)_0$ = 0.3, $(d\phi/d\sigma)_0$ = 3.17. Adopted from Prasanna (2017).

B_0 is the value at $r = r_0$ on the equatorial plane $\theta = \frac{\pi}{2}$. The poloidal components are the same as found by Prasanna and Verma (1977)

$$B_r = F_{\theta\varphi} = -(3\mu Sin\theta Cos\theta/4m^3)\left\{r^2 \ln\left(1 - \frac{2m}{r}\right) + 2m(r+m)\right\}$$

and

$$B_\varrho = F_{\varphi r} = ((3\mu \sin^2\theta/2m^2)\left(1 - \frac{2m}{r}\right)^{-1}\left\{(1 - m/r) + (r/2m - 1)\ln(1 - 2m/r)\right\}.$$

Prasanna and Sengupta have analysed the particle trajectories in the presence of toroidal fields and suggest that the main impact of the toroidal component is to rebound the particles in jet-like straight-line trajectories. Some typical plots are presented in figures (10.9). which are the projections on the $(x-y)$ and $(x-z)$ planes

From the single particle orbits it appears that as the particles approach the perihelion of the central body, depending upon their energy and angular momentum they get bounced off, particularly near the polar region as the magnetic field strength increases. Those which are directed away from their initial position $\left(\left(\frac{d\rho}{d\sigma}\right)_0 > 0\right)$ seem to spiral around before shooting away to infinity. Having seen that the pure toroidal field deflects the particle away from the central source while the poloidal field confines the particle it is necessary to look at the scenario when both the components are present. It appears (perhaps naturally) that the particle behaviour depends upon the relative strengths of the two fields. When $\lambda_{pol} \gg \lambda_{tor}$, particularly for black-hole (exterior) the particle gets sucked in while for neutron stars presence of the toroidal component yields, outgoing jet-like trajectories, particularly near the polar regions (Fig. 10.10).

Thus, detailed studies of charged particle motion for various space times with different geometrical features like for example static to slowly rotating blackholes with different superimposed magnetic fields (mostly dipolar) have revealed different types of orbits that could be helpful while discussing accretion disks both thin and thick as will be considered in the following.

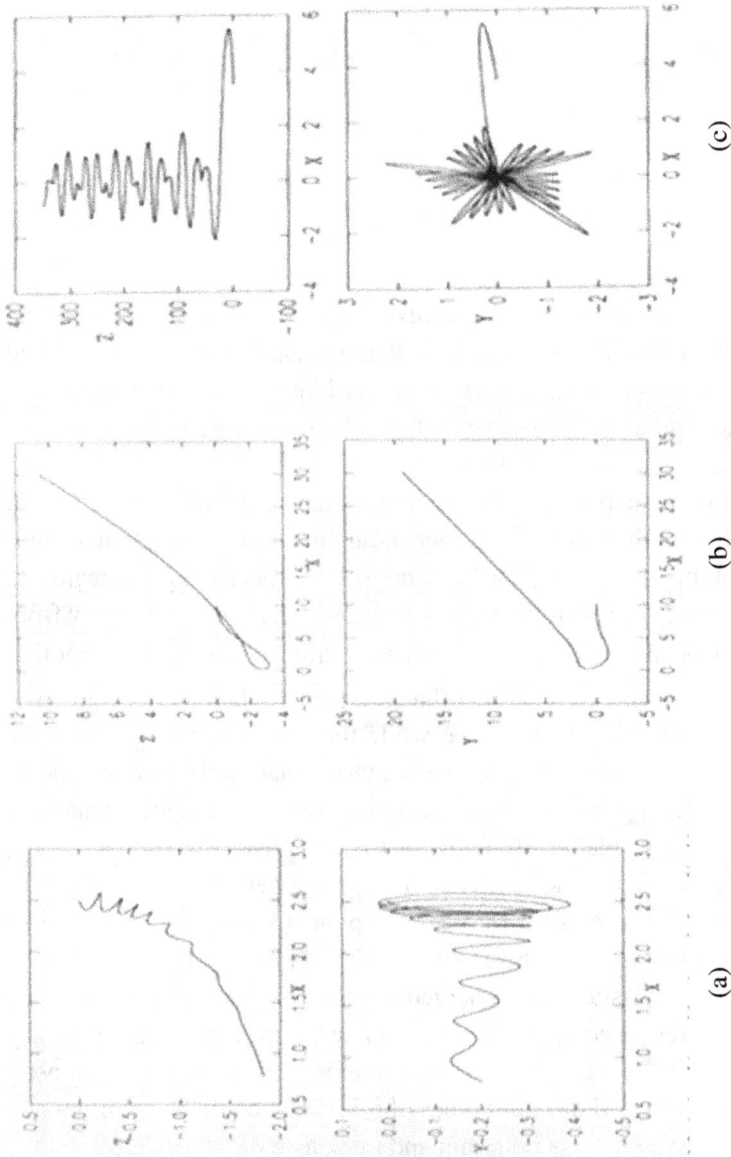

Fig. 10.10. Projections of the particle trajectories in a combined toroidal (λ_t) + poloidal (λ_p) magnetic field. (a) $E = 2$, $L = 91.76$, $\lambda_t = 8$, $\lambda_p = 80$, $\rho_0 = 2.5$, $(d\theta/d\sigma)_0 = 0.3$, $(d\rho/d\sigma)_0 = -1.92$. (b) $E = 5$, $L = 9.4$, $\lambda_t = 4$, $\lambda_p = 80$, $\rho_0 = 10$, $(d\theta/d\sigma)_0 = 0$, $(d\rho/d\sigma)_0 = -4.91$. (c) $E = 5$, $L = 41.38$, $\lambda_t = 400$, $\lambda_p = 80$, $\rho_0 = 3.5$, $(d\rho/d\sigma)_0 = 0.3$, $(d\theta/d\sigma)_0 = 4.9$. Adopted from Prasanna and Sengupta (1994) and Prasanna (2017).

Chapter 11

Accretion Disk Dynamics

In Chapter 7, while introducing the concept of accretion, it was mentioned that there are two types of accretion, spherical accretion and disk accretion. The discussion considered above is mostly relevant to spherical accretion where the accreting matter is devoid of angular momentum. However, while considering the particle orbits we have seen how the presence of angular momentum for an incoming particle gives the particle inbound orbits sometimes gyrating (charged particle in a magnetic field) otherwise circular around the compact object. In that case, the particles follow Keplerian orbits which can be stable. If such particles gather together they form a disk-like structure and if the matter has to accrete onto the central star, the particles should lose angular momentum making the particle spiral-in towards the star. The inflowing matter mostly in the case of binary stars could be neutral gas or charged fluid. Studying the dynamics of such flows one needs either ordinary fluid mechanics or magneto hydrodynamics both in Newtonian and Relativistic formalisms. These concepts were introduced in Chapter 8. There can be situations when the accreting matter is plasma (a combination of neutrals, ions and electrons) for which some of the concepts and formalisms were introduced in Chapter 9.

As is known, while Newtonian theory of gravity succeeds in explaining most of the aspects of physics up to the range of our solar system, general relativity goes beyond, explaining the finer features of the solar system like the perihelion advance and light bending, and going far deeper into the cosmos bringing out the aspects concerning the compact objects, black holes and several other cosmic features, particularly the strong field effects. There have been several studies of accretion physics dealing with active galactic nuclei and X-ray sources reviews of which may be found in Lightman *et al.* (1978), Pringle (1981) Wiita (1985), Begelman *et al.* (1984) and Frank *et al.* (1985). However, the earliest of the theoretical models that took into account various physical parameters self-consistently was due to Shakura and Sunyaev (1973) which came to be known as the standard model. This model is also known as the α-model as it assumes that the transverse stress in the disk is proportional to the total pressure through the relation $t_{r\varphi} = \alpha p$, α being a dimensionless parameter <1. The most important aspect of rotating matter accretion is its angular momentum transfer which is caused by the viscous stresses due to the friction between the layers of inflowing matter, and as a byproduct, the energy also is dissipated as heat. Friction not only causes the angular momentum to transfer outwards and mass inwards but also converts the gravitational energy of matter into heat. So produced heat radiation diffuses towards the top and bottom layers of the flow and gets radiated away. If the central star has a rigid surface like in the case of white dwarfs and neutron stars a shock gets formed close to it whereas in the case of black holes, the flow gets supersonic in the inner parts of the disk. While in the case of neutron stars or white dwarfs, the total luminosity does not depend upon the nature of the flow or radius, giving the value proportional to $(GM\dot{M})/R$, R being the radius of the star, in the case of blackholes the luminosity can depend upon the flow geometry as the efficiency for disk accretion being 0.057 for non rotating black holes, while it goes upto 0.42 for Kerr black holes. Following the discussion given by Treves, Maraschi and Abramowicz (1989), we consider briefly the analysis for thin disk dynamics as follows. Using the cylindrical polar coordinates with the flow being stationary and axisymmetric and assuming that the matter is confined to the plane z = 0, and the geometrical half thickness of the flow $H = H(r)$, the surface density of the disk is given by $\Sigma(r) = \int_0^H \rho(r,z)dz$, $H \ll r$, (thin disk approximation).With Ω representing the angular velocity the azimuthal velocity $v_\varphi = \Omega r$ is given by the pure Keplerian velocity $v_K = \sqrt{(GM/r)}$. The horizontal velocity v_r is purely subsonic as given by $v_r \ll v_s = \sqrt{(\partial p/\partial \rho)}$. The vertical component of the velocity is assumed to be negligible, $v_z < v_r < v_\varphi$, an approximation that accompanies the thin disk approximation. If p_r, p_g

represent the radiative pressure and the gas pressure, and τ^{es}, τ^{ff} represent the opacities due to electron scattering and free-free absorption, the disk may be divided into three regions.

(i) the inner region, where $p_r \gg p_g$ and $\tau^{es} \gg \tau^{ff}$,
(ii) the middle region with $p_r \ll p_g$ and $\tau^{es} \gg \tau^{ff}$ and
(iii) the outer region with $p_r \ll p_g$ and $\tau^{es} \ll \tau^{ff}$.

Along with these, it is also assumed that the pressure gradient $\partial p/\partial r$ is negligible. As the disk is in a steady state, the conservation of mass and angular momentum gives

$$\dot{M} = -2\pi r \Sigma v_r, \quad \text{and} \quad 2\pi r^3 \Omega \Sigma v_r = G(r) + C. \tag{11.1}$$

$G(r)$ represents the 'viscous torque' equal to $2\pi r^3 \nu \Sigma (\partial \Omega/\partial r)$ with ν representing the 'kinematic viscosity'. The constant C is related to the rate at which the angular momentum flows onto the central body which is determined by the condition that the viscous torque is zero at the inner edge. This gives a relation for the kinematic viscosity ν for Keplerian angular momentum distribution, the expression $\nu = (\dot{M}/3\pi\Sigma)(1 - R_{in}/R)^{1/2}$. With such a disk structure if one considers a cylindrical surface $R = R_0$, crossing the flow, the rates of mass and angular momentum flow across R_0 would be related as given by

$$\dot{J}(R_0) = \dot{M}(R_0)l(R_0) - G(R_0), \tag{11.2}$$

where $l(R_0)$ is the specific angular momentum showing that the total angular momentum $\dot{J}$ gets split into the advective part (due to macroscopic flow) and $\dot{M}l$ the viscous torque G representing the chaotic microscopic motion. According to Raleigh criterion, when $d\Omega/dr$ is positive, viscous torque transports the angular momentum outwards and for a stationary disk both $\dot{J}$ and $\dot{M}$ are fixed, the inner boundary condition $G(R_0) = 0$, yields

$$\dot{M}[l(R) - l(R_{in})] = 2\pi R^2 \nu \Sigma (d\Omega/dr), \tag{11.3}$$

which for the Keplerian angular momentum distribution gives with

$$f= [l(R) - l(R_{in})]/l(R) = (1-\sqrt(R_{in}/R)), \quad \dot{M}f = 3\pi\nu\Sigma, \tag{11.4}$$

as was found earlier. As was mentioned in chapter 8, in the context of hydrodynamics, to discuss the accretion flow with dissipative forces, it is necessary to recall the definitions of the coefficients of bulk viscosity ζ and shear viscosity μ which together are needed to consider the shear tensor which could dissipate the kinetic energy converting it to heat. The shear tensor is defined as

$$\sigma_{ik} = \zeta\theta\delta_{ik} + \mu\left(v_{i,k} + v_{k,i} - \frac{2}{3}\delta_{ik}\theta\right), \quad \theta = v^i v_{,i}. \tag{11.5}$$

μ is also referred to as dynamical viscosity as it is related to the kinematic viscosity ν through the relation $\mu = \nu\rho$. With these, the viscous shear tensor can be expressed as

$$t_{ij} = -2\int_0^H \sigma_{ij}\, dz. \tag{11.6}$$

In the present case, there is only one non-zero component

$$t_{r\varphi} = -2H\sigma_{r\varphi} = -2\mu HR\, d\Omega/dr = 3\mu\Omega H. \tag{11.7}$$

As viscosity is characterised mainly by the small-scale chaotic motion (turbulence) the coefficient ν can be defined as $\nu = \bar{v}_{turb} l_{turb}$, where $\bar{v}_{turb}$ is the average velocity of chaotic motion relative to the mean gas motion and l_{turb} is the average size of the turbulent cell. It is known that whenever the motion is supersonic, shocks will dissipate turbulent kinetic energy into heat and thus one requires $\bar{v}_{turb} \le v_s$ the local sound speed. Further, the cell sizes are bounded by the thickness of the disk $l_{turb} < H/3$. From these considerations, one finds the stress tensor to be

$$t_{r\varphi} \approx 3\rho\nu H\Omega \le \rho v_s H^2 \Omega \approx v_s \Sigma\Omega H. \tag{11.8}$$

As the disk structure is characterised mainly by the vertical hydrostatic balance, the pressure gradient along z being equal to the gravitational force along z gives the equation,

$$\frac{1}{\rho}dp/dz = -(GM/R^3)z = -\Omega^2 z. \tag{11.9}$$

Integrating and using the boundary conditions, $p = p_c$ along $z = 0$ and $p = 0$ along $z = H$, one finds $P = p_c (1 - z^2/H^2)$ with $p_c = \varrho\Omega^2 H^2/2$. Thus, one has $P = \Sigma\Omega^2 H/3$, which is equal to ϱv_s^2. Using this in the expression for the stress tensor yields $t_{r\varphi} = \alpha P H$, $\alpha \le 1$. This constant parameter α, which is always ≤ 1, for standard disk models is estimated depending upon the physical system and is generally in the range $0.01 < \alpha \le 1$.

11.1 Dissipation of Energy

One can find the expression for the total energy dissipation due to viscosity, which goes out as heat from either surface of the disk, (as given by Landau & Lifshitz 1989) to be

$$E_k = \varrho/2 \quad \int v^2 dv. \tag{11.10}$$

The Navier–Stokes equation in a slightly different notation from (8.8) may be expressed as

$$\partial v_i/\partial t = -v_j \partial v_i/\partial x^j - 1/\rho\,(\partial p/\partial x^i - \partial\sigma_{ij}/\partial x^j). \tag{11.11}$$

Taking the time derivative of (11.10) and using (11.11) one finds

$$\dot{E}_k = -\rho(v.\nabla)(v^2/2 + p/\rho) + \nabla\cdot[v^i\,(\partial\sigma_{ij}/\partial x^j)] - \sigma_{ij}(\partial v^i/\partial x^j). \tag{11.12}$$

For an incompressible fluid as *div v* = 0, the first term on the right-hand side can be rewritten as a total divergence thus getting the equation

$$\dot{E}_k = -\nabla\cdot[\rho v(v^2/2 + p/\rho)[v^i\,(\partial\sigma_{ij}/\partial x^j)]] - \sigma_{ij}(\partial v^i/\partial x^j) \tag{11.13}$$

with the first term representing the advective part of the energy flux density (due to the actual transfer of fluid) while the second term represents the internal friction of the energy flux. Integrating over a volume dV and then using Gauss theorem, one finds

$$\dot{E}_k = -\oint[\rho v(v^2/2 + p/\rho) - v\cdot\sigma]dS - \int\sigma_{ik}[\partial v^i/\partial x^k]dV. \tag{11.14}$$

Extending the integral over the entire volume of fluid, one gets (as the surface term goes to zero because the velocity at infinity is zero)

$$\dot{E}_k = \int \sigma_{ik} [\partial v^i / \partial x^k + \partial v^k / \partial x^i] dV = -1/2\eta^{-1} \int \sigma^{ik} \sigma_{ik} dv. \qquad (11.15)$$

Using the earlier derived expression for the stress tensor, and simplifying one gets

$$\dot{E}_k = -\rho \nu \pi R \ (Rd\Omega/dr)^2 \ dR \ dz. \qquad (11.16)$$

The surface heat generation due to viscosity as given by

$$Q^+(R) = \frac{1}{2\pi R} \frac{d}{dR} \left(|\dot{E}_k|\right) = (Rd\Omega/dr)^2 / 2 \int \rho v \, dz = \frac{\nu \Sigma}{2} (Rd\Omega/dr)^2 \qquad (11.17)$$

may be rewritten as

$$Q^+ = (\dot{M}/4\pi R) \ (l(R) - l(R_{in})) \ (d\Omega/dR) \qquad (11.18)$$

that finally gives for Keplerian angular momentum distribution

$$Q^+(R) = f \, 3GM\dot{M}/8\pi R^3, \quad f = 1 - \sqrt{(R_{in}/R)} \qquad (11.19)$$

and the disk luminosity defined as $L_d = \int_{R_{in}}^{\infty} Q^+ R dR$ turns out to be $GM\dot{M}/2R_{in}$ which is one-half of accretion luminosity. Thus one has for *stationary disks the total luminosity given by the product of the accretion rate $\dot{M}$ and the binding energy of the particle in its last stable circular orbit R_{in} and independent of the dissipative terms.*

With steady accretion disks, the total energy balance is governed by the set of equations (Shakura & Sunyayev 1976):

$$\frac{1}{\rho} \frac{dp}{dz} = -(GM/R^2)z = \Omega^2 z, \qquad (11.20)$$

for the vertical hydrostatic equilibrium,

$$q(z) = -(c/3\rho\bar{\sigma})(dE_r/dz) \qquad (11.21)$$

for the radiative transfer, and the equation of total energy balance being given as

$$dq/dz = (3/8\pi)(GM\rho/R^3)(\dot{M}/E)\left(1 - \left(\frac{R_{in}}{R}\right)^{\frac{1}{2}}\right) = Q^+/\Sigma \qquad (11.22)$$

with E_r and $\bar{\sigma}$ representing the radiation energy density and opacity of matter in the disk respectively. Considering the vertical hydrostatic balance equation with approximation $\partial P/\partial z \approx P/H$ and using the equation for average pressure $\rho\Omega^2 H^2/3$, one finds $H \sim v_s/v_\varphi R$, implying $v_\varphi \gg v_s$. If the disk fluid is isothermal, then in the vertical direction one has

$$\rho(R,z) = \rho_c(R) \exp(-z^2/2H^2), \tag{11.23}$$

and the pressure P is given by the sum of the gas pressure and the radiation pressure, which for ionised hydrogen plasma under local thermo-dynamical equilibrium is $P(\rho,T) = \frac{2\rho kT}{m_p} + \frac{a}{3}T^4$.

As the main agency for heat transport is radiation the temperature T must be given by the energy balance equation relating the energy flux in the vertical direction to the rate of energy generation by viscous dissipation. If the gas is optically thick ($\tau \gg 1$) then the radiative transfer equation may be solved under diffusion approximation. This gives the flux of radiant energy through a surface z-constant, to be (Shapiro and Teukolsky, 1983)

$$F(R,z) = -c/3\frac{d}{d\tau}(aT^4), \quad \tau = \bar{\kappa}(\varrho,T)\Sigma, \tag{11.24}$$

$\bar{\kappa}$ being the total Rosseland mean opacity. Integrating the above over the half thickness of the disk one gets

$$\int_0^H F(R,z) = Q^+ = caT^4/3\tau,$$

which finally gives the expression for T, to be

$$(4\sigma/3\tau)T^4 = (3GM\dot{M}/8\pi R^2)[1-(R_{in}/R)^{\frac{1}{2}}] \tag{11.25}$$

With these, the set of equations that govern a steady-state thin accretion disk are

$$\Sigma = H\varrho, \quad H = \frac{v_s}{\Omega}, \quad P = \rho v_s^2,$$

$$P = 2\frac{\rho kT}{m_p} + \frac{aT^4}{3}, \quad t_{r\varphi} = \alpha PH, \quad \alpha \le 1, \quad \tau = \kappa_R(\rho,T)\Sigma,$$

$$\nu\Sigma=(\dot{M}/3\pi)[1-(R_{in}/R)^{\frac{1}{2}}],$$

$$T=(9\tau GM\dot{M}/32\pi\sigma R^3)[1-(R_{in}/R)^{\frac{1}{2}}]^{1/4}. \tag{11.26}$$

Given $M,\dot{M},R$ and α one can then determine all other parameters of the disk. Shkura and Sunyaev (1973) were the first to obtain a complete solution for such a disk. An important aspect to notice from their solution is that the luminosity of the disk goes inversely to the nearness of the inner edge of the disk (R_{in}), while the temperature goes as $(1 - (R_{in}/R)^{1/2})$. *This in a sense indicates that the closer the inner edge of the disk is to the compact object, the energy release is more.*

As the inner regions of the disk are likely to be dominated by the radiation pressure it is necessary to consider whether there are any constraints regarding the thin disk approximation $H \ll R$. Given that $P_r \gg P_g$ and the opacity is mainly due to electron scattering ($\tau = \rho H\sigma_T/m_p$, σ_T being the Thomson scattering cross section)

$$P=a\frac{T^4}{3}\equiv4\bar{\sigma}T^4/3c=(3GM\dot{M}\sigma_T H\rho/8\pi R^3 m_p c)f.$$

As $P/\rho=v_s^2=H^2\Omega^2=H^2GM/R^3$, one finds from the above equation the scale height to be

$$H=(3\dot{M}\sigma_T/8\pi m_p c)[(1-(R_{in}/R)^{1/2}].$$

As radiation pressure is limited by the Eddington luminosity $\dot{M}_c=4\pi GMm_p/c\sigma_T$, one can express the scale height through the equation $H = \left(3\varepsilon R_{in}/2\right)(\dot{M}/\dot{M}_c)f$, ε being the efficiency factor equal to GM/Rc^2. This suggests that the thin disk approximation is valid only for the accretion rate $\dot{M}\ll\dot{M}_c$. For a thin but optically thick ($\tau \approx 1$) disk, the surface temperature is given by the effective black body temperature

$$T\approx10^7(M/M_{sun})^{(-1/2)}(\dot{M}/\dot{M}_c)^{1/4}(R_s/R)^{3/4}f^{1/4}$$

which attains maximum at $R = 49/36\,R_{in}$, where $dQ^+/dR = 0$. one thus finds

$$T_{max}=5\times10^6(M/M_{sun})^{(-1/2)}(\dot{M}/\dot{M}_c)^{1/4}(R_{sun}/R_{in})^{3/4}.$$

If $R_{in} = 6\,R_{sun}$, for a 10 M_{sun} black hole, $T_{max} \approx 10^6\,K$, while for a 10^8 M_{sun} black hole $T_{max} \approx 10^3\,K$. one can thus expect x-ray photons from a thin disk around 10 M_{sun} black hole, while in the case of a $10^8\,M_{sun}$ black hole one can expect optical to UV photons. The following diagrams depict the equatorial temperature gradient (Fig. 11.1) and the surface density gradient (Fig. 11.2) for the cases of two black holes mentioned above (Treves *et al.*, 1989).

The study of the stability of thin accretion disks was carried out by several authors (Piran, 1978; Lightman, 1974; Pringle *et al.*, 1973; Shakura & Sunyayev (1976); Shibazaki and Hoshi, 1976) and it was generally found that the inner regions of the disk are secularly and thermally unstable. It is possible that the nature of the equations of state produces a non-linear behaviour which could be causing the instability.

Figure 11.3 depicts the behaviour of $\dot{M}$ as a function of Σ the surface density, as it exhibits non-linearity which could show that the stability could be affected by the change of slopes of the curves. However, this

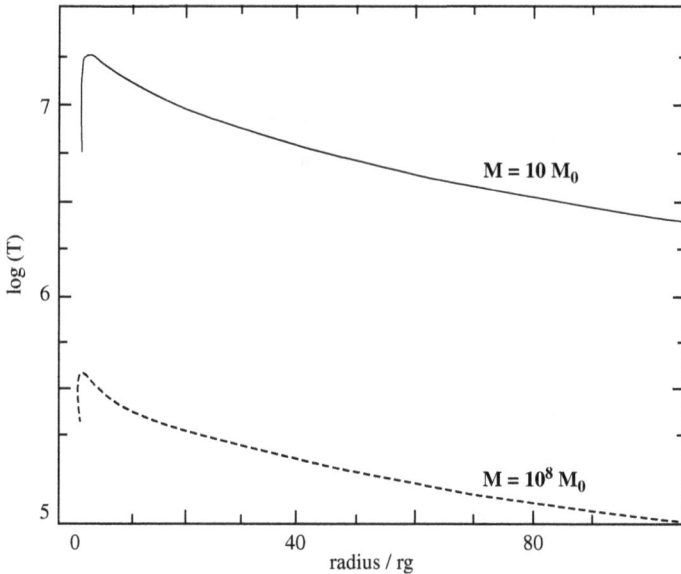

Fig. 11.1.

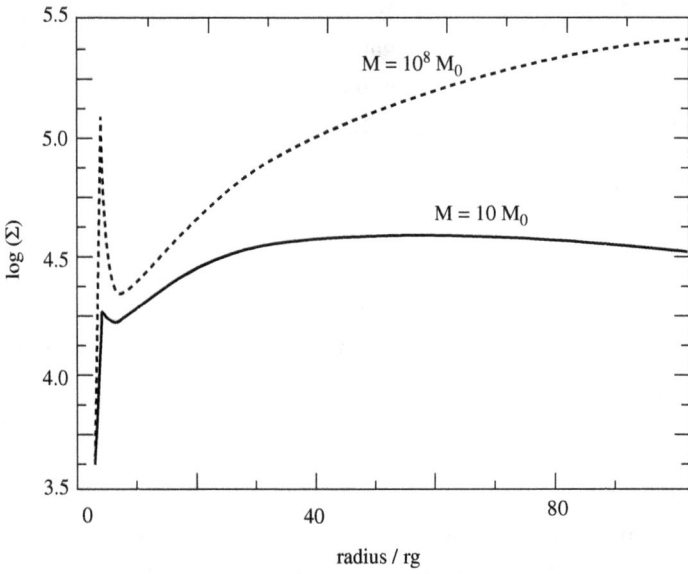

Fig. 11.2.

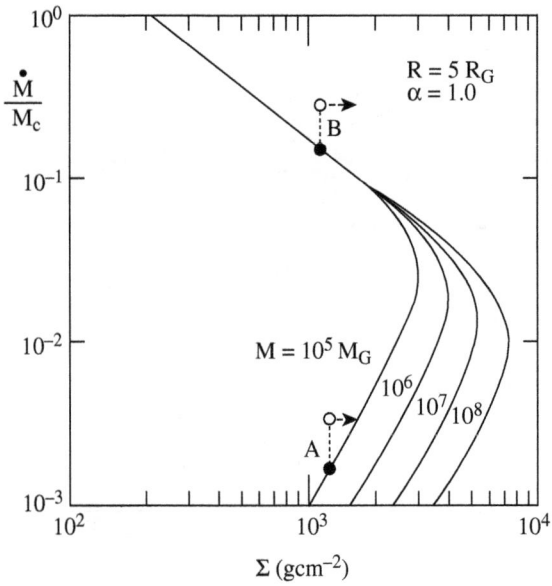

Fig. 11.3.

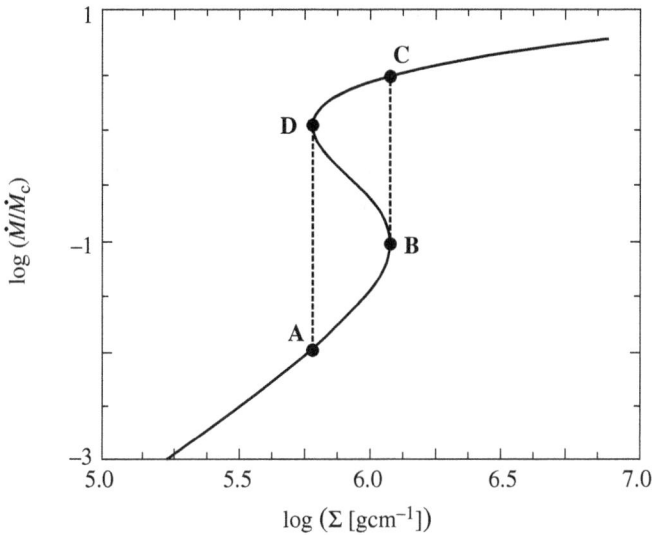

Fig. 11.4.

feature became visible in the work of Abramowicz and Marsi (1987) showing the existence of a limit cycle (Fig. 11.4) as the curve $\dot{M}$ (Σ) that could give a bend which can give another stable branch. As argued by Maraschi *et al.* (1987) for such accretion rates ($\dot{M} \approx \dot{M}_c$) Shakura Sunyaev model may not be adequate as one may have to include both horizontal pressure gradient and heat transport. $\dot{M}(\Sigma)$ curve is characteristically *S*-shaped (Fig. 11.4) where the upper and lower branches refer to stable disk models and the middle one to unstable ones. If the accretion rate is such that the model lies in the region as marked in the Fig. 11.4 (AB) then stationary accretion would not be possible.

As is seen in Fig. 11.4 for higher accretion rates all the different curves converge to a unique line as there is no mass dependence asymptotically. The turning points for all these curves occur always for $\beta = 2/5$, and $(d \ln \Sigma/d \ln M) = (5\beta - 2)/(2 + 3\beta)$, where β is the ratio of total pressure to gas pressure.

If the accretion rate $\dot{M}$ exceeds the critical value $\dot{M}_c$ leading to super Eddington luminosity, then the radiation pressure dominates such that the inner regions of the disk gets blown up ($H \approx R$) and further the horizontal gradient of pressure will get to be non-zero. In such a case, if the radial drift is still negligible, then the angular momentum distribution will no longer be Keplerian. Such disks are called **thick accretion** disks that are supposed to play significant role in the modelling of Quasars.

Chapter 12

Thick Accretion Disks

In the case of thick disks, as the disk equations show when the accretion rate reaches the limiting value that leads to Eddington luminosity, the scale height of the disk H becomes comparable to the radius R_{in} thus making the disk geometrically thick. As Wiita (1985) points out, in thick disks, pressure and centrifugal forces are of comparable magnitude. This makes even for sub-critical accretion rate, the clump instabilities that affect the thin disks could lead to time averaged bloated structure in the inner regions. The vertical structure of the disk would look like a torus and in some cases resemble a sphere with two deep narrow funnels along the rotation axis Fig. 12.1.

The structure within the disk depends upon whether the equilibrium is attained through the pressure balance between the pressure forces against gravity or the centrifugal force. As Prasanna and Chakraborty (1980) point out the formation of a cusp at the inner edge is revealed only when the structure is described through the inclusion of the general relativistic analysis and not in the Newtonian formulation. Also as mentioned by Abramowicz *et al.* (1978) apart from showing that the cusp exists between r_{mb} and r_s where the marginally bound and stable time-like trajectories appear, it has also been pointed out that there exist five different possibilities regarding the disc structure depending upon the angular momentum distribution as follows:

THICK ACCRETION DISKS

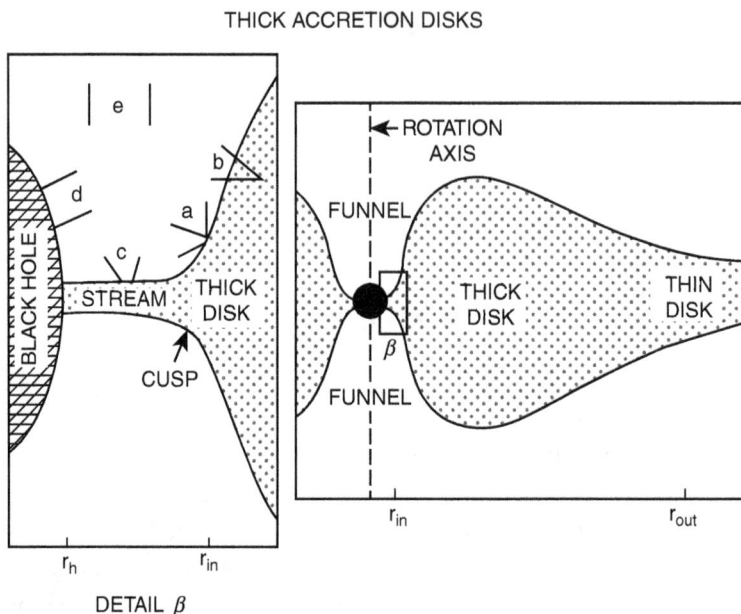

Fig. 12.1.   Adapted from ACN 80.

(1) $l_0 < l_{ms}$ disk will not form,

(2) $l_0 = l_{ms}$ disk exists as an infinitesimally thin, unstable ring located on the circle $r = r_{ms}$,

(3) $l_{ms} < l_0 < l_{mb}$ many disks can form without cusp but only one with cusp.

(4) $l_0 = l_{mb}$ disk exists with a cusp located on the marginally closed equipotential surface, and

(5) $l_0 > l_{mb}$ disk has no cusp.

l_0, l_{ms}, l_{mb} denote the initial angular momentum, and angular momenta at the stable orbit and marginally bound orbits respectively. As shown by Seguin (1975) the existence of the cusp for any stable angular momentum distribution depends on the fact that the angular momentum is increasing outwards that can allow the distribution to cross the Keplerian angular momentum, only at two points. In the case of a Schwarzschild black hole the cusp is located at the sonic point which occurs between r_s and r_{mb}. If $\dot{M} \gg \dot{M}_c$ the cusp and the sonic point coincide at r_s and if otherwise the cusp moves towards r_{mb}. As the energy per particle released in the accretion process is the binding energy at the last stable orbit coinciding with the cusp a stationary disk with $\dot{M} \gg \dot{M}_c$ has lower efficiency compared to the standard disk as the binding energy at r_{mb} tends to zero.

12.1 Perfect Fluid Disk on Schwarzschild Geometry

As a simple example, Prasanna and Chakraborty (1980) have considered the structure and stability of an accretion disk rotating around a static black hole, under axisymmetric perturbations, whose exterior is described by the Schwarzschild geometry. For the case of an incompressible ($\rho = const$) fluid, the pressure and velocity profiles are as follows: In geometrical units

$$p = \rho\,[\mathrm{B}\{(1-2/R)^{-1} - A/R^2\sin^2\theta\}^{1/2}-1], \; v = A(1-2/R)/R^2\sin\,\theta, \quad (12.1)$$

the constants A and B as given by

$$A = 2\,(ab)^2/(a+b)(a-2)(b-2), \text{ and} \qquad (12.2)$$
$$B = [(b^2 - a^2)(b-2)(a-2)/(b^3(a-2) - a^3(b-2))]^{1/2}$$

with $R = r/m$, $a = r_d/m$, and $b = r_b/m$ and a constraint that the pressure p > 0, throughout the interior and goes to zero at the boundary, leading to the condition that the inner edge cannot lie within $r = 4m$ and further $b > 2a/(a-4)$, a lies between 4 and 6. There are no restrictions when $a \geq 6$, for b. r_a, r_b denote the radii where the inner and outer edges intersect the plane $\theta = \pi/2$. In the meridional plane the condition for the disk edge θ_e is given by

$$\sin^2\theta_e = AB^2 \Big/ \left[R^2\,(B^2\left(1 - \frac{2}{R}\right)^{-1} - 1)\right]. \qquad (12.3)$$

In order to compare with the non-relativistic case one can take the Newtonian limit of the governing equations. Fig. 12.2 shows the

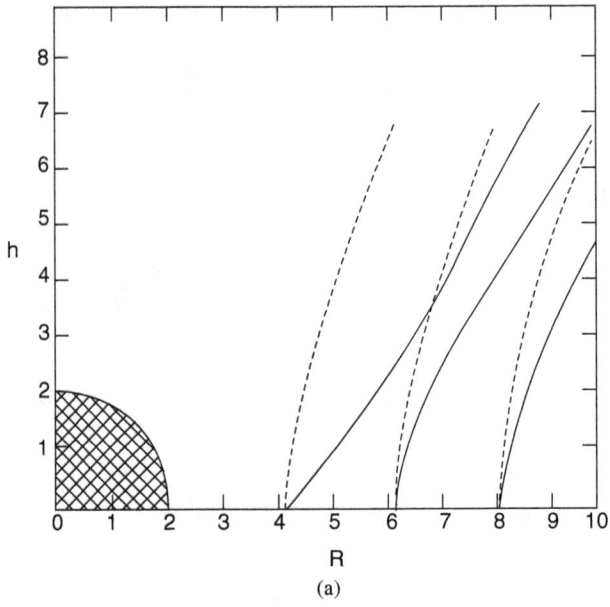

(a)

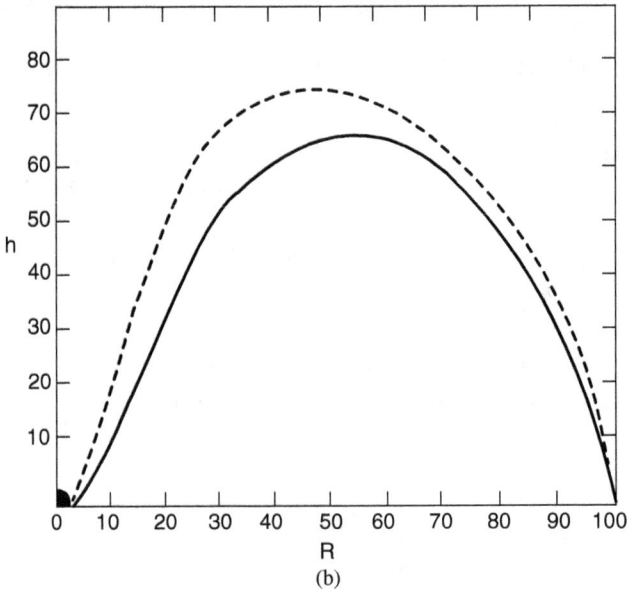

(b)

Fig. 12.2.   Adopted from (PC 82).

meridional cross-section of the disk both for the Newtonian (a) and general relativistic (b) formulations.

As can be seen, the important difference between the two formulations is shown regarding the cusp formation at the inner edge. This result matches with that of (Kozlowski *et al.*, 1978) who show that the inner edge should be at a distance greater than 4m as the last bound orbit for particles exist at $r = 4m$ in Schwarzschild geometry. As Abramowicz *et al.* (1980) also point out in the case of thick disks for accretion onto black holes a self-crossing equipotential surface (Roche lobe) having the shape of a cusp is realised mainly after taking into account the general relativistic modification to the Newtonian dynamics.

12.2 Accretion Disks Around Compact Objects with Self Consistent Electromagnetic Fields

As the complete set of equations for such a system consists of the Einstein-Maxwell equations along with the Maxwell equations on curved spacetime, a coupled set of second-order partial differential equations governing the various physical parameters, one will have to resort to some simplification through a possible set of assumptions which can decouple the system. The set of assumptions are

  (i) the disk is not massive in comparison with the central source (black hole) and thus the background space-time geometry is entirely governed by the central object,
 (ii) the associated electromagnetic field also is of very small energy that does not affect the geometry, but do get modified by the background gravity, and
(iii) the disk is in equilibrium under the action of gravitational, centrifugal, pressure gradient and electromagnetic forces.

Considering the disk to be made up of charged perfect fluid with pressure p, matter density ρ, charge density ϵ, electrical conductivity σ, the laws of conservation of energy-momentum may be expressed as $T^{ij}_{;j} = 0$, where the stress-energy tensor is given by

$$T_i^j = \left(p + \frac{\rho}{c^2}\right)u_i u^j - \frac{p}{c^2}\delta_i^j - \frac{4\pi}{c}E_i^j, \tag{12.4}$$

and the electromagnetic stress-energy tensor by $E_i^j = F_{ik}F^{jk} - \delta_i^j F_{kl}F^{kl}$, with the field tensor F^{ij} satisfying Maxwell's equations $F_{\cdot j}^{ij} = J^i$, and $F_{(ij,k)} = 0$. along with the generalised Ohm's law $J^i = c\varepsilon u^i + \sigma F^{ij}u_j$. For details of the calculations, one may look at the reference (Prasanna 1984). If we consider the fluid without viscous terms then the equations for mass and momentum conservations are given by

$$\rho_{,j}u^j + \left(\rho + \frac{p}{c^2}\right)u_{,j}^j = \frac{1}{c^3}F_{ik}J^k u^i, \text{ and} \tag{12.5}$$

$$\left(\rho + \frac{p}{c^2}\right)u_{,j}^i u^j - \left(\frac{p_{,j}}{c^2}\right)h^{ij} = \frac{1}{c^3}[(F_k^i - F_{lk}u^i u^l)]^k, \tag{12.6}$$

where $h^{ij} = g^{ij} \pm u^i u^j$, the projection tensor, the choice of sign depending upon the signature of the metric. As it is very difficult to consider the entire system of governing equations one goes either for numerical methods or to try the analytical method after assuming several plausible assumptions concerning the disk parameters. In the case of numerical solutions attempts have been made to write three-dimensional magneto-hydrodynamical codes for simulation of structures of such disks. As this path had been pursued by several authors since the eighties and has become almost an industry, some authors found it more interesting to follow the analytical solutions path after making different assumptions with respect to physical parameters and their relations for plausible situations. In this section, we shall consider a few of these approaches and analysis.

12.3 Disks with Magnetic Fields

One of the earliest discussions that mention qualitatively the influence of magnetic fields on the inner edge of the disk is due to Pringle and Rees (1972) while discussing accretion as a source for compact X-ray sources. They obtain limits on the accretion radius and a lower bound for the disk density depending upon the magnetic field strength of the accreting neutron star. As Lightman *et al.* (1978) point out at high temperatures attained close to the compact object the particle mean free paths are so long that a fluid dynamical treatment will not be self-consistent unless collective effects get operative due to interstellar magnetic fields. These magnetic fields even if initially negligible, due to stretching of the field lines make the energy density vary as r^{-4}, and thus making it dynamically important. Bisnovati Kogan and Blinnikov (1972) and Ichimaru (1977) have considered the effects of magnetic field on the accreting plasma and have found that there could be an increase in the efficiency in radiation emission and that the turbulence gets generated mainly due to differential rotation of the plasma which decays through current dissipation due to anomalous magnetic viscosity. Galeev *et al.* (1979) using slightly more rigorous treatment for generating a magnetic field due to the differential motion of conductive media have found that the fastest reconnection mechanism is not sufficiently rapid to develop effectively in the inner portions of the disk. They have also pointed out that the build-up is limited by nonlinear effects due to convection. On the other hand, this analysis has revealed that the disk could develop a magnetically confined and structured carona consisting of many small scale extremely hot caronal loops that could emit both soft and hard X-ays depending upon the disk luminosity. Ghosh and Lamb (1979) had considered accretion by rotating magnetised neutron stars and had found some constraints on possible models and using the solution for a two-dimensional hydro-magnetic equations they calculated the torque on a magnetic neutron star accreting matter from a Keplerian disk. These investigations seem to have shown that there exists coupling between the central star and the inner boundary of the disk that leads to a conclusion that the spin-up torque is appreciably less for fast rotators than slower ones and further, there could be break in stellar rotation. Several studies by Prasanna *et al.* (Prasanna and Bhaskaran, 1989; Bhaskaran and Prasanna, 1990; Tripathy *et al.*, 1990; Bhaskaran *et al.*, 1990; Tripathy *et al.*, 1993), have shown that for a given angular momentum distribution the inner edge of the disk can reach well within the limit given by the last stable circular orbit if the surface magnetic field of the star is not high as well as the matter density at the outer edge and the accretion rates are within reasonable limits. In the case of accretion onto neutron stars, the rate of rotation of the star is important for getting an equilibrium structure for the disk and as was shown for the case of single particle orbits the momentum balance at the boundary layer is different for co- and counter-rotating disks (Prasanna and Dadhich, 1982). The role of the magnetic field in defining the boundary layer is important as the interaction of currents within and outside the disk could have different strengths for co- and counter-rotating plasma disks. Some analysis of the stability under radial perturbations of the disk at the inner boundary, which includes the effect of finite conductivity in the disk dynamics, within the local approximation seems to show that the disk is stable to Kelvin–Helmholtz and resistive electromagnetic modes whereas the magnetosonic mode can destabilise the disk structure. In the context of disks with magnetic fields, most of the studies have considered only a dipolar field. However, Begelman *et al.* (1984) have considered possible effects when there is also a toroidal component arising mainly from the backward-bent field lines tied to the gas due to inertia. This toroidal field might be helpful in collimating the hydromagnetic out flow forming jets as the hydromagnetic stresses can fling the gas outward through the centrifugal force. The toroidal component of the magnetic field will have associated hoop stress which can help in collimation of the disk material (Blandford *et al.*, 1990).

A magneto-fluid thick disk, stationary and axisymmetric with finite conductivity (σ) around a compact object is considered in Newtonian analysis along with Maxwell's equations and Ohm's law, and the fluid equations are solved analytically and a class of solutions for different magnetic field components are looked into. The azimuthal current produced due to the motion of the magnetofluid modifies the magnetic field structure inside the disk. The self-consistent pressure profiles of the thick disk show that the magnetic moment associated with the central compact object is constrained by the prescribed value of density at the outer boundary and finite conductivity. For higher values of the angular momentum, the pressure profiles exhibit a turning point indicating the possible existence of instabilities in the disk configuration (Tripathy *et al.* 1990). The relevant equations for this analysis may be found in Prasanna (2017). The momentum equations together give the Bernoulli equation

$$d\tilde{p} + \rho\left[d\left(\frac{v^2}{2} - \frac{MG}{r} \right) \right] = 0, \quad \tilde{p} = p + B^2/8\pi. \tag{12.7}$$

For a thin disk confined to the plane $\theta = \pi/2$, the condition $v^\theta = 0$, $v^r \ll c$, puts a lower limit on σ and the density and pressure of such a disk are given by

$$\rho = 4\pi\sigma \dot{M}/(1-k)c^2 r,$$

$$p = p_0 + 4\pi GM\dot{M}\sigma/(1-k)c^2 r^2 - \frac{\dot{M}}{3r^3}\{K+L^2/K\} - B_1^2/8\pi. \quad (12.8)$$

$$K = (1-k)c^2/4\pi\sigma$$

The integration constants are determined using proper boundary conditions. Following Fishbone and Moncrief (1976) if one chooses $L^2 = nGM/r_{in}$, one finds when the pressure inside the disk is positive, one can constrain n to range from 0 to 1.5. It is interesting to note that even within this range, depending upon the value of k the constant associated with the electromagnetic field, there exists a critical n for a given k beyond which the pressure profiles change from decreasing with r to increasing which means, the solution has changed from disk type to wind type. The following figures give the plots for typical pressure profiles for a few cases of thin disks. Fig. 12.1.

Pressure profiles for a thin plasma disk.

(a) $\rho_{out} = 10^{-4}$ gms/cc, $B_0 = 7.10^8 G$, $N = 12$, $x_{in} = 15$ m;
(b) k = 0, (c) N = 6, $x_m = 9m$ adopted from (Tripathy *et al.*, 1990)

In the case of thick disks, for incompressible fluid, one finds from the Bernoulli equation, $p = \rho_0(p_0 + GM/r - v^2/2) - B^2/8\pi$. The continuity

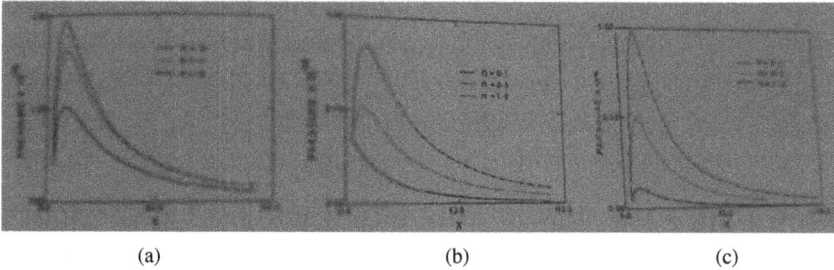

(a)   (b)   (c)

Fig. 12.3.

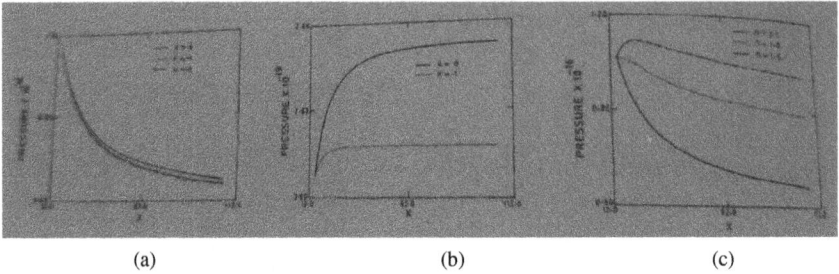

(a)   (b)   (c)

Fig. 12.4.   Adopted from Tripathy *et al.* (1990).

equation in this case may be expressed as *div v* = 0, which gives *v* = *curl*
F, with F being arbitrary. In a simple case the velocity equations may be
solved (TPD90) and the solution expressed as,

$$v^r = A\cot\theta/r - (1-k)c^2/4\pi\sigma r, \quad v^\theta = -\frac{A}{r} - (1-k)c^2\cot\theta/4\pi\sigma r. \quad (12.9)$$

The boundary condition that $v^r\left(r, \frac{\pi}{2} \pm \theta_1\right) = 0$, gives the constant A
to be equal to $\pm(1-k)c^2\cot\theta_1/4\pi\sigma$. The pressure profiles are depicted in
Fig. 12.4.

It is useful to note that the accreting plasma in the presence of a dipole
magnetic field gives rise to an azimuthal current and a non-zero charge
density. This current generates the disk field which is continuous across
the boundary due to the presence of finite resistivity of the plasma. As
there is no net flow of angular momentum an axisymmetric thick disk can
exist only in the steady state. The pressure profiles show that for the different strengths of the magnetic fields there exist both upper and lower
bounds for the compact object for which the disk equilibrium is meaningful. Again the presence of large pressure gradients indicates the possible
occurrence of plasma instabilities.

12.4 Disk with Dipole Magnetic Field on Linearised Kerr Geometry.

It is useful to look at the pressure profiles for thin disks around slowly
rotating compact objects to get some idea about the influence of the

rotation of the central star on the disk structure and physical parameters keeping some of the parameters fixed. Let us consider the case of a slowly rotating compact object, with the Kerr parameter $\alpha = \frac{a}{m} \ll 1$, and an axisymmetric stationary disk around it which has only the azimuthal component of the 3-velocity non-zero. The plasma in the disk is assumed to be of infinite conductivity and the associated magnetic field has only poloidal components non-zero. The space-time geometry can be assumed to be represented by the linearised Kerr geometry. The equations of motion as well as the Maxwell's equations are as given in (Bhaskaran *et al.*, 1990, eqns. 2.1–2.9) The absence of the φ component of $(B \times J)$ is consistent with J^r and J^θ being zero and one finds the azimuthal magnetic field component to be $B_\varphi = \frac{K}{1-\frac{2m}{r}} \sin\theta$. The constant K may be chosen to be zero and using the force-free condition (generalised Ohm's law), one can write the poloidal components of the electric field to be: $E_r = V^\varphi B_\theta / c$ and $E_\theta = -V^\varphi B_r / c$. One can assume that the azimuthal velocity to be relativistic Keplerian, modified by the inertial frame dragging due to the rotation of the central source as given by

$$v^\varphi = L\left(1 - \frac{2m}{r}\right)/r^2\sin^2\theta - 2amc/r^3\sin^2\theta. \qquad (12.10)$$

When the relevant equations are expressed in a locally non-rotating frame (LNRF) defined in (4.6.2), the pressure profiles are as given in Fig. (12.5.) for disks of constant density $\rho = 10^{-8}$, $B_0 = 10^9 G$ and $l = 1$, (a,b,c) $\rho = 10^{-7}$, $B_0 = 10^{10}G$ and $l = 1$, (d,e,f).

As can be expected, in the case of accretion onto neutron stars, both rotation and the strength of the surface magnetic field play significant roles. First thing to notice is that the momentum balance at the boundary layer between the star and the disk's inner edge shows different features for co- and counter-rotating disks. If the radial velocity was to be non-zero, the boundary layer could exhibit more sensitive features due to the interaction of currents within and outside for varying strengths of co- and counter-rotating disks.

So far one considered only the influence of the poloidal magnetic field on disk equilibrium. As was shown in the context of single particle motion (Prasanna and Sengupta, 1994) the presence of the toroidal component of the magnetic field does bring in some changes and as was pointed out by

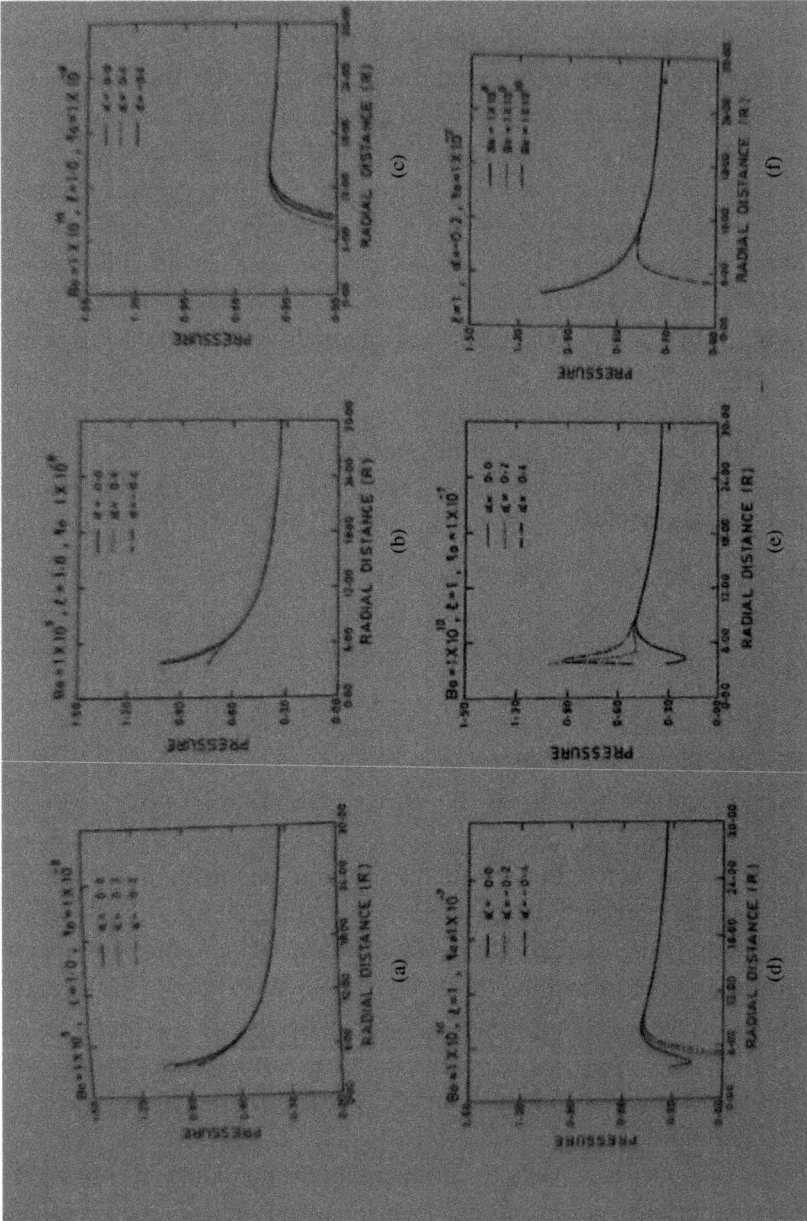

Fig. 12.5.

Blandford *et al.* (1990) the hoop stress that arises can help in collimation and acceleration of jets. As a preliminary study for this aspect, a non-accreting magnetofluid disk around a non-rotating central object which is axisymmetric and stationary has been considered in the Newtonian approximation for which the basic MHD equations are given by

$$\nabla \cdot (\rho_m \nu) = 0, \quad \rho_m(\nu \cdot \nabla)\nu = -\nabla p + \rho_m g + (J \times B)/c \qquad (12.11)$$

along with the Maxwell's equations and Ohm's law, where ρ_m is the matter density, ν the flow velocity and g the acceleration due to the gravity of the central compact object (Banerjee *et al.*, 1995). The equilibrium solutions obtained are physically plausible as the pressure is positive everywhere within the disk having a local maximum on the plane $\theta = \pi/2$, provided the two parameters α representing the ratio of gravitational energy to kinetic energy at $r = r_{in}$, and β the ratio of the toroidal field strength to the poloidal field strength satisfy the following inequalities:

$$\beta^2 < 1.87(\alpha - 1)^{-7/8}/|(\alpha - 3.5)|, \qquad (12.12)$$

$$\beta^2 < 3.67\alpha^{-15.8}, \qquad (12.13)$$

$$\text{and } \beta^2 < (105/224)\,(\alpha - 1)^{-15/8} \qquad (12.14)$$

In Figure 12.6, the shaded region depicts the parameter space allowed by these inequalities with α ranging between 1 and 3.5. For $\alpha \geq 1.1$, the strongest constraint comes from the last inequality as also the constraint that β must be less than 2.

The vertical (meridional) structure of the equilibrium plasma density $\rho_m(\theta)$ and of the pressure are expressed by

$$\rho_m(\theta) = (15/8)(\alpha - \sin^8\theta)^{-15/8} + \beta^2, \qquad (12.15)$$

$$p(\theta) = (15/8)(\alpha - \sin^8\theta)^{-7/8} + \beta^2\left(\alpha - \frac{7}{2}\sin^8\theta\right) \qquad (12.16)$$

which are presented in Figs. 12.7. to 12.10, the first two for the case $\beta = 0$, with no toroidal component and consequently no azimuthal current, showing that the external magnetic field has no role in defining the structure

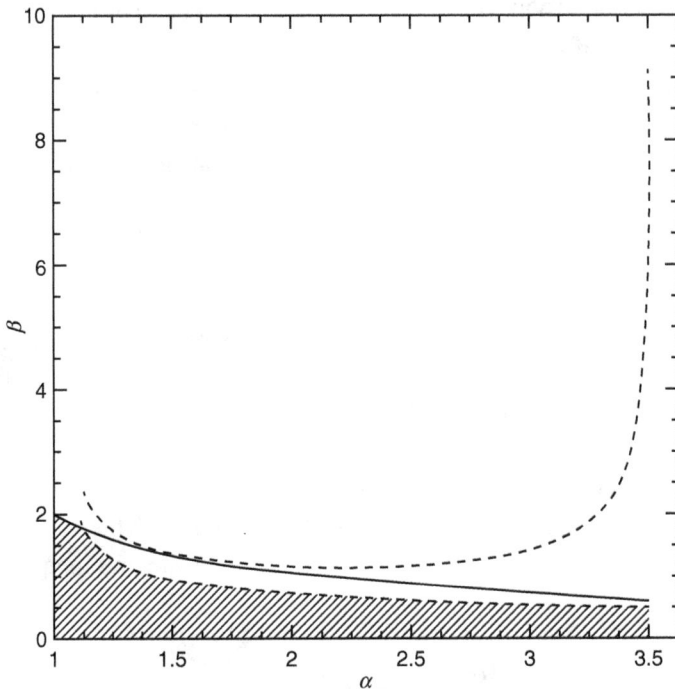

Fig. 12.6. (The parameter space (α, β). The area subtended below the dot-dashed curve shows the allowed parameter space described by the inequality (12.12) and the area below the solid line (12.13) and the area below the dashed line represents (12.14).

of the disk. Both pressure and density profiles show global maxima on either side of the plane $\theta = \pi/2$, for all values of α.

Further one can also see that as $\beta \to 1$, both profiles show sharp gradients in the meridional direction indicating that there would be a rapid decrease of the matter density away from the equatorial plane under equilibrium. This change could be a consequence of the fact that the toroidal magnetic field helps the pressure forces to counter balance the centrifugal force and thus making the pressure gradient less steep. After the value $\theta = \sin^{-1}\left(\frac{2\alpha}{7}\right)^{1/8}$ on either side of the equatorial plane the isotropic part of the pressure dominates over the plasma pressure with $\beta = 0$, while for higher values of $\beta \to 1$, there is a large gradient in the meridional direction. These features seem to exist for all allowed values of α and β as presented in Fig. 12.6.

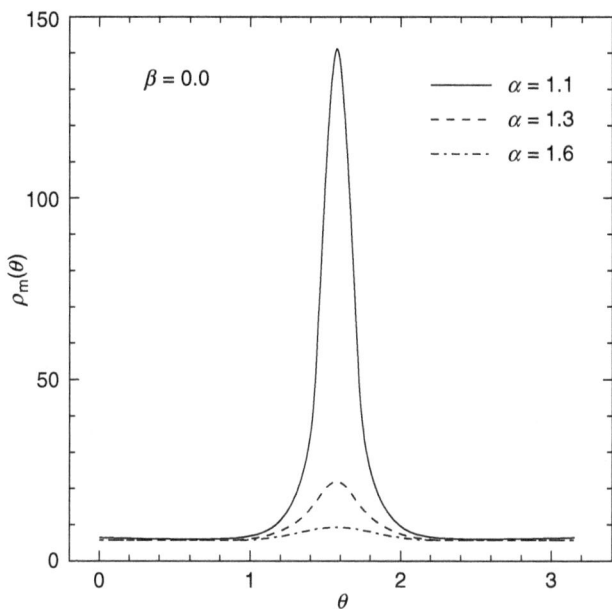

Fig. 12.7.

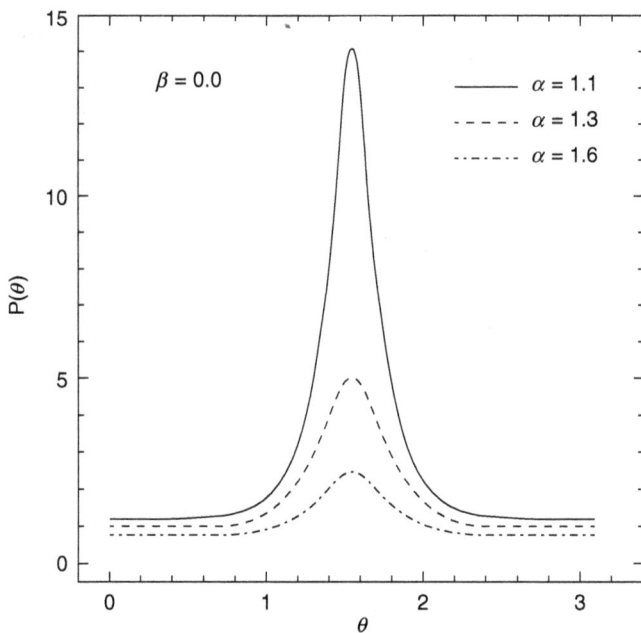

Fig. 12.8.

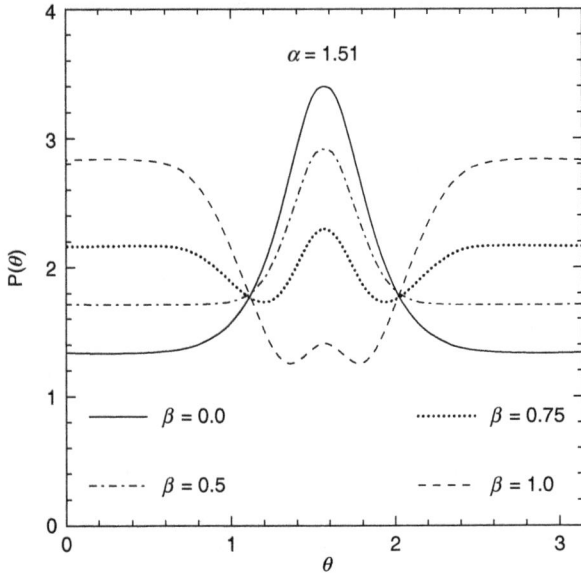

Fig. 12.9.

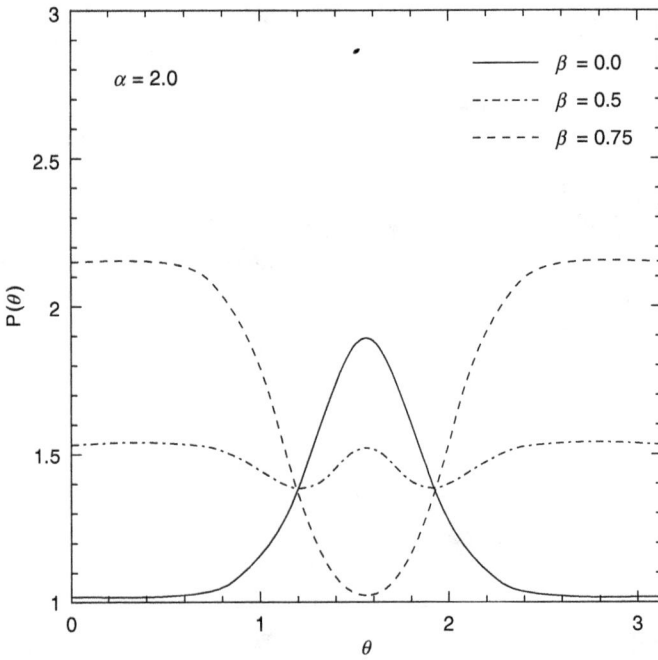

Fig. 12.10.

Figs. 12.11–12.15 depict the magnetic field topology

Projections of a magnetic line of force within the disk on the XY plane as a function of θ (0 to π) at r = 10m, R = 7m, for different vaues of B_1/B_0.

The toroidal component of the magnetic field and the poloidal component both satisfy the equation that their divergences are zero, with B_T = (0, 0, B_φ) and B_p = (B_r, B_θ, 0). From the given relevant equations one can find the field structure for different values of β. The above figures depict the projections of the field lines on the *X–Y*, *Y–Z*, and *X–Z* planes. For the case β = 0, as there is only a poloidal field the field line is depicted as a straight line in the *Y–Z* plane. When $\beta \neq 0$, the projection is a loop caused

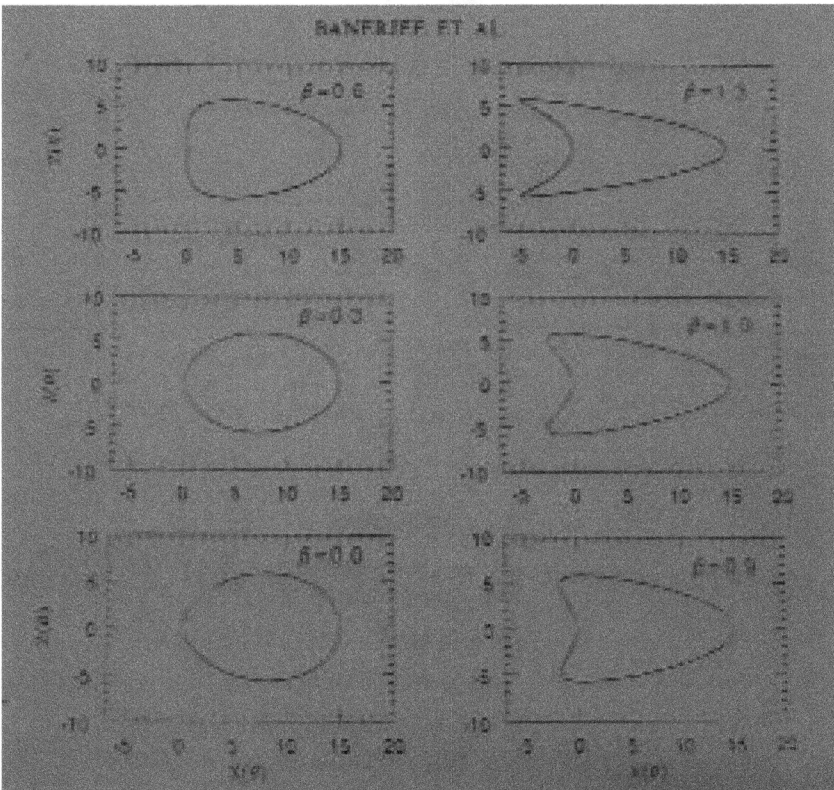

Fig. 12.11.

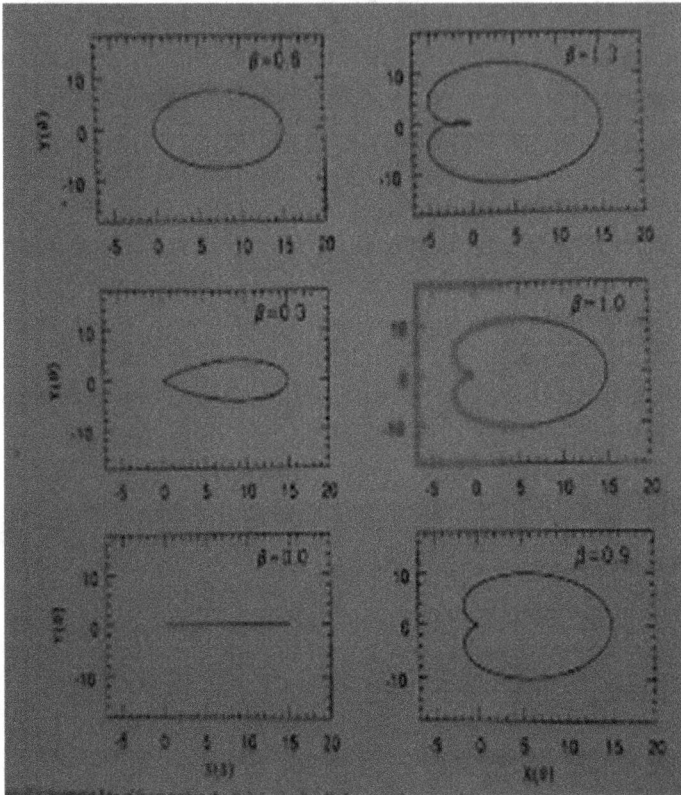

Fig. 12.12.

by the toroidal component and as the β value increases the line of force is stretched more with a size of the loop getting larger. To allow this possibility kinks develop. In this case, the toroidal field line also has a straight line projection in the X–Y plane perpendicular to the one in the Y–Z plane for a pure dipole field. This reflects in the appearance of two lobes on the Y–Z plane which grow in size as β increases and tilt in the horizontal direction. Finally as shown in the projection on the X–Z plane inclusion of the toroidal component deforms the loop by streching and compressing indicating a shear in the magnetic field lines as shown in Fig. 12.14. All these indicate that for a self-consistent magnetohydrodynamic thick disk where the rotation of the compact object with a poloidal field and a self-generated toroidal component produces shear kinks that could store sufficient energy

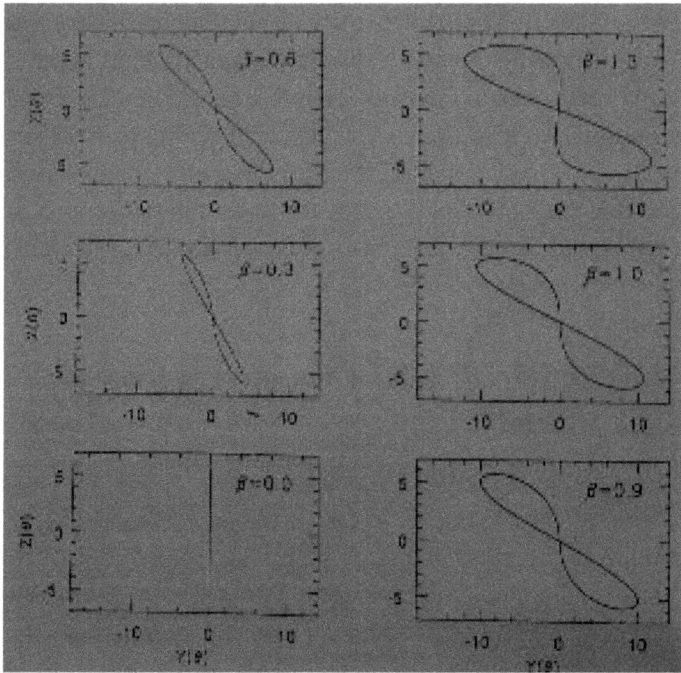

Fig. 12.13.   Projection on the YZ plane.

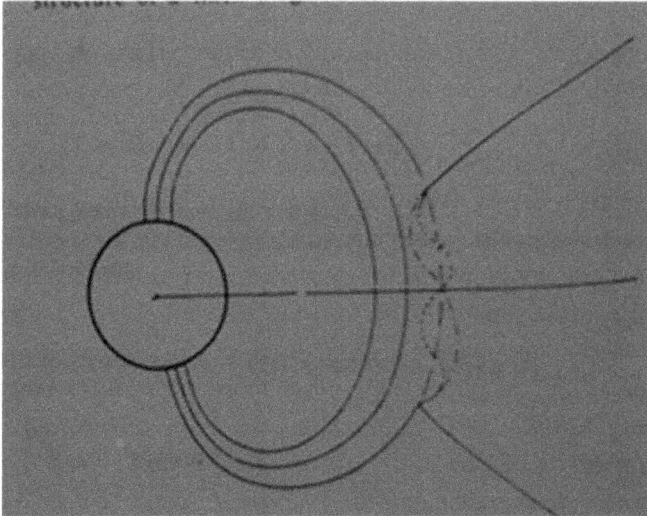

Fig. 12.14.   Schematic diagram of magnetic lines of force at the inner edge of the disk.

to generate instability. These kinks increase with increasing toroidal components, again showing the role of toroidal magnetic field in the study of thick plasma disks around compact objects with dipole magnetic field. It is important to take a look at these studies extended to general relativistic treatment as the gravitational field of the central source increases the strengths of the magnetic fields as already shown in the context of single particle orbits and also generates toroidal component self consistently (Prasanna and Sengupta, 1994).

12.5 Axisymmetric MHD Disk on a Schwarzschild Background

We next take up the discussion of an axisymmetric MHD disk structured around a non-rotating compact object with geometry as given by the Schwarzschild background geometry with a superposed dipole magnetic field. As has been pointed out by Znaek (1976) the strong magnetic field can bring the inner edge of the disk closer to the compact object as was also shown by Prasanna *et al.* while discussing the effective potential curves for a charged particle in such a painted dipole magnetic field. (Prasanna and Varma, 1977; Prasanna and Vishveswara, 1978). Further from physical considerations, as shown by Goldreich and Julian (1969) there can be significant plasma density near the surface of a neutron star and as a result of high magnetic fields processes like pair production also can play an important role and as indicated by Gonthier and Harding (1994) intense gravitational field near a compact object can also influence such physical processes. With these points in view, it becomes necessary to examine axisymmetric stationary solutions of the plasma (MHD) equilibrium in the corotation regime in the geometry of strong gravitation. Though the investigations of Balbus and Hawley (1991) point out that accretion disks with a weak magnetic field are unstable, as explained by Camenzind (2000) *the physical reason for this instability is fairly simple. Considering the above situation and the effect of perturbing a weak vertical field threading an otherwise uniform disk, if the field remains frozen in the plasma, field lines connecting adjacent annuli in the disk will be sheared by the differential rotation into a trailing spiral pattern. Provided the field is weak enough, magnetic tension will not keep the field lines* back to the vertical. *The magnetic tension acts to reduce the angular momentum of the inner fluid element and boost that to the outer one, providing angular momentum transport in the outward direction that is required to drive the accretion process'.* However as argued by Knobloch (1992) under axisymmetric shear perturbations, with the presence of the toroidal component such instabilities are absent. But as discussed in the previous section there always seems to exist a toroidal component of the magnetic field in the case of rotating compact objects with a dipole field coming from the bending of the dipole field lines because of their frozen in structure. It is remarked that the solution for velocity profiles satisfies Feraro's iso rotation law in the corresponding Newtonian analysis. As for most of the pulsars, the angular momentum parameter is sufficiently small (Gonthier and Harding, 1994) invoking the study in Schwarzschild geometry in general relativity seems quite satisfactory.

We first takeup the case that goes to quasi-Keplerian flow velocity in the flat space limit, where the pressure and enthalpy profiles do not satisfy the barotropic conditions even when the term $J \times B$ is not there. It is possible that this is so because when the flow velocity is quasi-Keplerian, the term $(\nu \cdot \nabla)(\nabla \times \nu)$ is non zero. Also, the inclusion of the $J \times B$ term can introduce additional non barotropicity. Further, the nonzero toroidal magnetic field introduces additional inhomogeneities in the pressure and enthalpy profiles. The strength of these inhomogeneities for $\beta \approx 1$ is of the same order as for equilibrium without any $J \times B$ force.

It has been noted that in order to keep the plasma pressure positive in the entire disk region, as also to have a global maximum for the pressure along the plane $\theta = \pi/2$, the parameters α and β which are now functions of the Schwarzschild coordinate r have to satisfy certain inequalities which in the flat space limit are same as the ones obtained in the Newtonian analysis, namely $\alpha > 1$ and $\beta \sim 2$. The inequalities referred to relate to the rotational state of the disk and the toroidal component of the magnetic field. In the relativistic formalism, the parameters are given by $\alpha_{min} = 1/a_1$ and

$$\beta_{max} = \sqrt{\left[\frac{3.67}{a_4} \left\{ \frac{7}{5} a_1 - \frac{2a_3}{5a_4} \right\}^{(-15/8)} a_1^{15/8} \right]}, \tag{12.17}$$

where the parameters a_1 to a_4 are defined in (Banerjee *et al.*, 1997). This parameter space is depicted in Figs. (12.15 and 12.16) along with the pressure profiles in the meridional plane and the profiles for α and β.

The parameter space, (α, β) at the Schwarzschild coordinate $\bar{r} = r/m = 6$, is shown for the case in which the plasma azimuthal velocity profile is quasi-Keplerian in the asymptotic limit. The area below the dashed and solid lines shows the domain of values for α, β allowed by the inequalities. (adopted from (BBDP 97) 12.17 shows the plot of β_{max} as a function $\bar{r} = r/m$, and the next the variation of pressure along the meridional direction as a function of θ, from 0 to π at $r = 6m$.

It has been found that the pressure and enthalpy profiles for quasi Keplerian velocity profile and the condition that pressure be positive with local maximum along the equatorial plane is maintained for all ranges of the parameters α and β unlike in the non-relativistic case. As nowhere

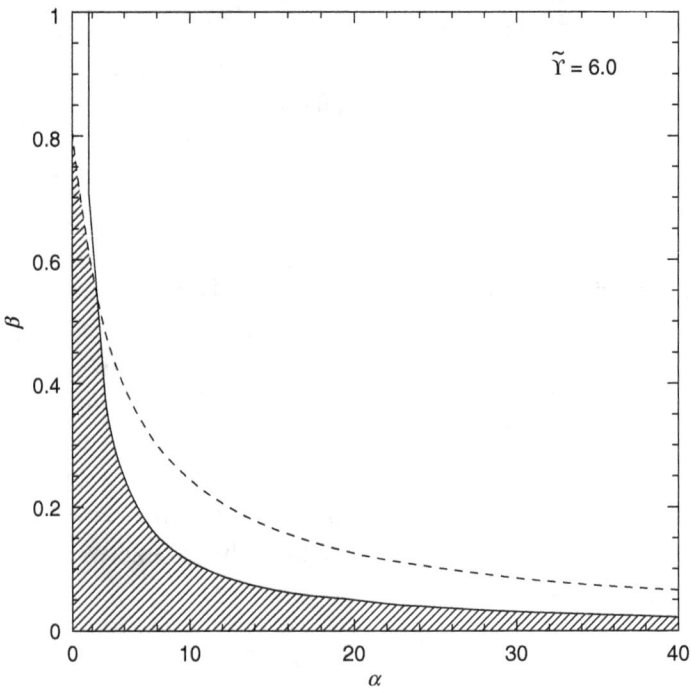

Fig. 12.15.

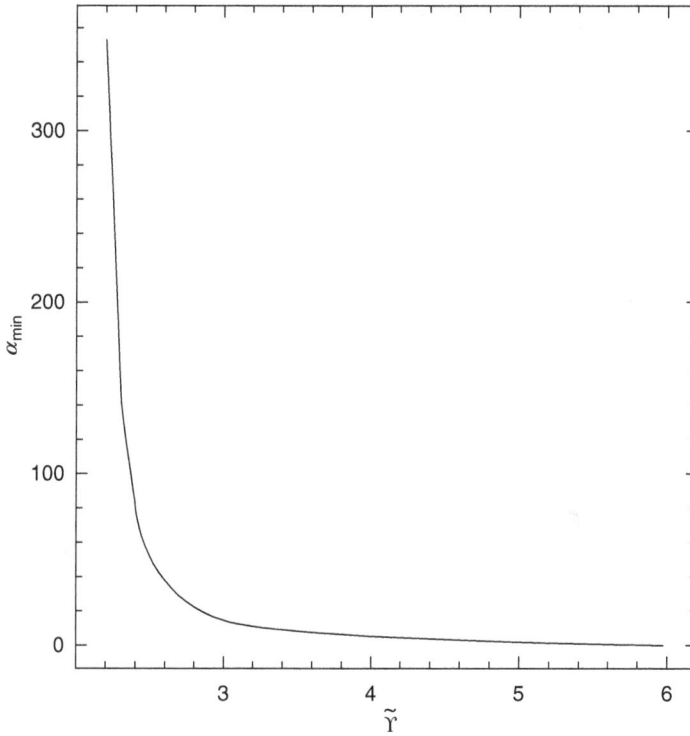

Fig. 12.16.

the thin disk approximation was made, the solution discussed can be used to study the magnetic torque exerted by a thick disk on the central star by incorporating the effects of accretion flow and finite resistivity perturbatively.

In the context of the study of rotating stars, it was pointed out how the rotation of the central star can influence the spacetime around it through the effect of inertial frame dragging that gives rise to gravimagneto effect. Extending this to disks around a rotating star, it has been found that physically meaningful thin disk solutions can exist wherein the angular momentum of the fluid in the disk arises due to the frame dragging induced by the central star (Prasanna, 1989). The equilibrium of the disk exists mainly due to the balancing of the hydrostatic gas pressure acting against the radiation pressure coming from accretion. If there is no bulk

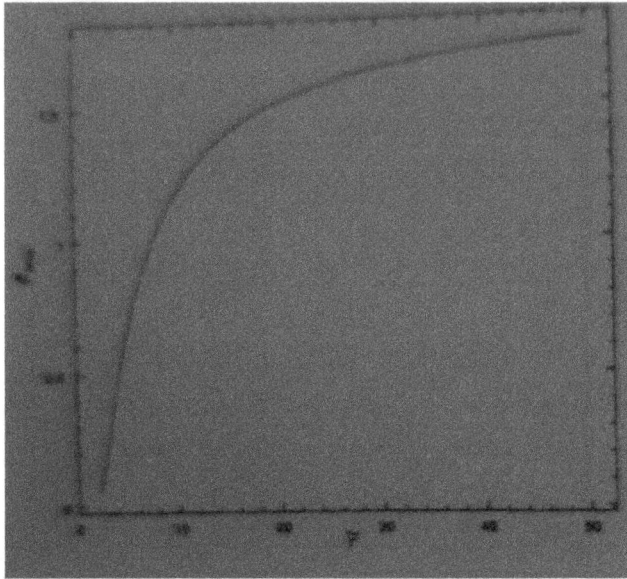

Fig. 12.17.

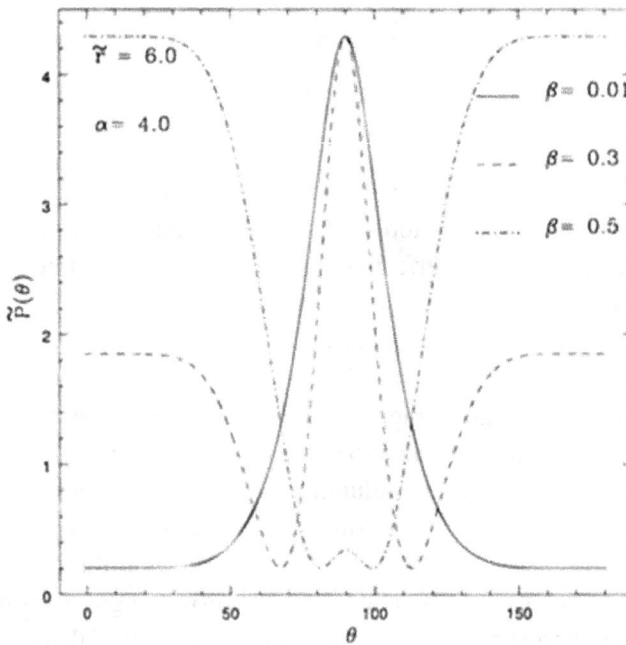

Fig. 12.18.

viscosity of the fluid then the pressure remains constant throughout the disk but with increasing bulk viscosity the pressure drops at the inner edge but soon stabilises to a constant value.

In view of this, it is important to reconsider the discussions of spherical accretion around rotating stars as the purely radial inflow at infinity would acquire angular momentum as it approaches the star and a corotating disk has to form before accretion which also requires that the approaching fluid needs to be viscous. If the radial velocity of the flow is high enough to overcome the dragging it could happen that the flow could get to the stage, particularly for super Eddington accretion rates, when there could be advection of heat inwards horizontally. Such a flow as discussed by Abramowicz *et al.* (1988) shows when $\dot{M}$ is plotted against surface density plane, a S-shaped curve together with two standard branches corresponding to spherically symmetric accretion models implying a limit cycle behaviour already referred to. Considering the roles of horizontal pressure and entropy gradients in moderate to super Eddington accretion rates it has been shown that they are important in the inner most transonic part of the disks orbiting around blackholes. Further attention has been drawn to the fact that the effect produced is a strong horizontal heat flux which changes the energy balance in the disk affecting the stability pattern wherein the well-known condition $p_g/(p_g + p_m) < 2/5$ seems to disappear for high enough $\dot{M}$. Such advection-dominated accretion flows (ADAF) have been discussed by Narayan *et al.* (1994, 1995) in sufficient details concluding that such flows have very low radiative efficiency as compared to gas flows in standard disks. Several authors have investigated the advection-dominated models and it seems that if the infalling gas has low optical depth and low density (as $\dot{M}$ becomes very small) the radiation time scale becomes lower than the accretion time scale as a result of which almost all the internal energy is lost to the hole. Having given a formalism they have dealt with self-similar solutions (which appear to be closer to some global solutions) that are valid away from the boundaries and almost up to the marginally stable orbit, but do not have transonic regions. It is possible that in some realistic situations, the radial velocity could approach the value c as one goes closer to the event horizon and this could indicate the existence of a sonic point. However, it may be noted that the self-similarity would break down if the deviation from the pseudo-Newtonian potential or the specific angular momentum j accreted by the black hole is greater or of the order ΩR^2, Ω being the azimuthal velocity of the accreting gas and R is the radial coordinate as shown by the numerical solutions of Gammie and Popham (1998). Bhatt and Prasanna (2000) introducing the effect of pseudo-Newtonian potential perturbatively demonstrate that this approach can give better insight into the parameter space of the solution by indicating that the self-similarity can be good over a large range of the viscosity parameter. However for certain values of the viscosity parameter α and the gas parameter γ the perturbations can have a singularity indicating the violation of self-similarity at distances far away from the black hole. Further, as they summarise the magnitude of the coefficient of perturbation in sound velocity remains less than one and consequently the ratio of perturbation to the unperturbed value remains finite over the entire range of R, $R_g < R < \infty$, showing that the self-similarity is a valid approximation. The same result is reflected in the ratio of the perturbed to unperturbed values of v_0 which decreases with increasing α. It appears that the possible violation of self similarity occurs mostly in the low α regime. Few more details on this discussion may be found in Prasanna (2017).

12.6 Inertial Forces and Accretion

Before ending the discussion on the accretion disks it may be useful to briefly look into the role of Coriolis force which in fact is quite important while discussing the physics of fluids around a rotating compact object. Though in general relativity particularly for blackholes, one finds that all aspects of rotation get combined in the role of the Kerr parameter, it has been found that with a special slicing of the four spaces into 3 + 1 structure using what is called 'optical reference geometry' (Abramowicz *et al.*, 1988) with conformal slicing some useful new features can be analysed. One finds particularly in a stationary space-time, which has a timelike Killing vector, one can with a conformal reslicing obtain a 3 + 1 splitting of the spacetime, where the 3-space is the quotient space obtained from the timelike Killing vector and the metric conformal to the spatial geometry of the original four-space. Such a splitting was first proposed by Carter *et al.* (Abramowicz *et al.*, 1988) and one is referred to original article for details. A covariant formalism of this splitting was made by Abramowicz *et al.* (1993, 1995) which brought out the exact expressions for the four accelerations in terms of the Newtonian concept of gravitational, centrifugal and Coriolis accelerations. Though this approach did lead to some special features (not realised in full GR) like centrifugal reversal, (Abramowicz and Prasanna, 1990) and a maximum ellipticity for a rotating body (Abramowicz and Miller, 1990) we shall not deal with them here and refer the interested reader to the original and follow up literature on these aspects. Applying these features of inertial forces one finds the split up of the total force in a general axisymmetric, stationary spacetime to be as given by

$$(F_g)_i = -\nabla_i \, \phi = \frac{1}{2} \partial_i \left\{ \ln \left[\frac{(g_{t\varphi}^2 - g_{tt} g_{\varphi\varphi})\}}{g_{\varphi\varphi}} \right] \right\}, \qquad (12.18)$$

$$(F_{Co})_i = -A^2 (\Omega - \omega) \sqrt{g_{\varphi\varphi}} \left[\partial_i (g_{t\varphi} / \sqrt{g_{\varphi\varphi}}) + \omega \partial_i \left(\sqrt{g_{\varphi\varphi}} \right) \right], \qquad (12.19)$$

$$(F_{cf})_i = -(A^2 / 2)(\Omega - \omega)^2 g_{\varphi\varphi} \partial_i \{ \ln[g_{\varphi\varphi}^2 / (g_{t\varphi}^2 - g_{tt} g_{\varphi\varphi})] \}. \qquad (12.20)$$

With $A^2 = -[\eta^i \eta_i + 2\Omega \eta^i \xi_i + \Omega^2 \xi_i \xi_i]$, ξ and η being the associated space-like and time-like Killing vectors. For the case of Schwarzschild geometry, these force components are given by

$$(F_g)_r = \frac{m}{r^2} \bigg/ \left(1 - \frac{2m}{r} \right), \quad (F_{Co})_i = 0, \qquad (12.21)$$

$$(F_{cf})_r = \frac{m_0 l^2}{r^3} (1 - 3m/r) \{ [r^2 (r - 2m)/(r^3 - l^2 r + 2ml^2)] \}. \qquad (12.22)$$

It can be noted immediately that outside the horizon r > 2m, the centrifugal force reverses its sign at r = 3m, the location of the last photon orbit (Abramowicz and Prasanna). This feature has some interesting consequences as noted below.

1. Raleigh criterion and viscous torque:
Consider a collection of particles orbiting a central gravitating source in Keplerian orbits (circular orbits) having angular momentum $l = l \, (r)$

satisfying the equilibrium condition $F(r, l^2) + f(r)$, $f(r)$ being all other position-depenent forces acting on the system, with the centrifugal force alone being dependent on angular momentum. Displacing a small element of the collection from the radius r_0 to $r_1 = r_0 + \delta r$, there will be an unbalanced force as given by

$$\Delta F = [F(r_1 l_0^2) + f(r_1)] - [F(r_1 l_1^2) + f(r_1)] = -(\partial F/\partial l^2)(dl^2/dr)\delta r \quad (12.23)$$

Now for the stability of the system, one should have $\Delta F \, \delta r < 0$ implying $(\partial F/\partial l^2)(dl^2/dr) > 0$. In Newtonian theory as both these terms are positive the stability is assured with angular momentum being transferred outwards. On the other hand here one finds the condition

$$\frac{m_0}{r}\left(1 - \frac{2m}{r}\right)\left(1 - \frac{3m}{r}\right)/\left[r - l^2\left(1 - \frac{2m}{r}\right)\right]^2 \}(dl^2/dr) > 0. \quad (12.24)$$

which requires outside the horizon $(r > 2m)$, $r > 3m$ for $(dl^2/dr) > 0$, and $r < 3m$ for $(dl^2/dr) < 0$. *This means that the Raleigh criterion for stability of the fluid configuration reverses sign at the last photon orbit* $r = 3m$. This particular analysis helps in clearly understanding the paradoxical feature as had been found by Anderson and Lemos (1988) where they argued that the commonly assumed zero torque boundary condition at the last stable Keplerian orbit is not tenable. The viscous torque is nonzero both at the last stable orbit and on the horizon. The angular momentum flux through a spacelike surface S is given by $j = \dot{M}l - Q$, where $\dot{M}l$ is the net mass flux multiplied by the specific angular momentum and thus is the advective part which is macroscopic while Q is the viscous torque the microscopic transport of angular momentum generally defined as $Q = -2\int \eta \sigma_{ij} \xi^j dS^i$, with σ_{ij} being the shear tensor and ξ^j the space like Killing vector and dS^i an oriented element of the surface and η the coefficient of viscosity. For the present case $\sigma_{ij}\xi^j$ is given by $A\Omega \, (1-3m/r)/\phi^{1/2}$, ϕ being the gravitational potential which clearly shows that the viscous torque changes sign at the last photon orbit $r = 3m$, there by confirming the claim that the advection of angular momentum is inwards closer to the horizon.

12.7 Effect of the Inertial Forces on Accretion Disk Dynamics

Discussing the roles of inertial forces, it is important to look at the role of the Coriolis force which in general relativity appears in the context of the 'frame dragging' for a rotating body. As we have now a special framework for splitting the four forces into separate units in a 3 + 1 splitting of the four space the role of individual forces makes for interesting discussion. Further as mentioned earlier, for discussing fully relativistic dynamics as one resorts to a numerical approach it would be important to relate the usual ADM splitting (which forms the base for numerical solutions) of the four-space to the Optical reference geometry parameters which might be useful for numerical procedures. On a general manifold defined by the metric $ds^2 = g_{ij}dx^i dx^j$, the (3 + 1)ADM splitting is defined through the lapse function α, shift vector β^a and the three metric γ_{ab} and the metric

$$ds^2 = -(\alpha^2 - \beta_a\beta^a)dt^2 + 2\beta_a dx^a dt + \gamma_{ab}dx^a dx^b \qquad (12.25)$$

wherein the indices $i,j,...$ go from 0 to 3 while a,b ... go from 1 to 3. As pointed out in Prasanna(Prasanna 2002) for numerical hydrodynamics, with the introduction of a unit time like vector field n^i normal to the hyper-surface Σ (t = constant) one can define the fluid three velocities (York, 1983) $V^a = U^a/\alpha U^t + \beta^a/\alpha$, U^a being the spatial component of the four vector $U^i = (U^a, U^t)$. Without going into the details of calculations (one can refer to the original) it may be found that the components of the inertial accelerations may now be written as

$$cf_a = \gamma^2\left[\left(VV^b\partial_b(V_a/V)+(V_aV^b\partial_b - V^2\partial_a\right)\phi - \frac{1}{2}V^bV^c\partial_a\gamma_{bc}\right] \qquad (12.26)$$

$$Co_a = -\left(\frac{\gamma^2}{\alpha}\right)\left[V\beta^b\partial_b(V_a/V)+V^b\partial_a(g_{0b})-\beta^cV^b\partial_a(\gamma_{cb})\right] \qquad (12.27)$$

As both forces depend explicitly on the 3-velocity one needs to solve the fluid equations with a given equation of state and the velocity vector so obtained may be used to check the features. For the details, one may refer to (Mukhopadhyay and Prasanna 2003). It has been pointed out that the centrifugal reversal occurs only through the dominance of the azimuthal velocity over the radial velocity. Whereas in a static spacetime, the results are the same as for a collection of circular geodesics, in the case of incompressible fluid (ρ = const) it is noted that at $r = 3m$ for $l = 3\sqrt{3}m$ the 3-velocity tends to the light velocity and thus requires to set l less than this value. The reversal then occurs at two values of r, one close to the photon orbit and the other closer to the horizon depending upon the choice of the angular momentum component. In principle, the expression representing the centrifugal acceleration has a quinttic equation in r and thus possesses five locations which depend upon the constants chosen (Mukhopadhyay and Prasanna, 2003). However only two of the five lie inside the photon orbit at $r = 3m$. In the case of Kerr blackholes due to the inherent property of frame dragging which effectively increases the angular velocity, the reversal can occur even for dusty fluids. When the pressure is non-zero the gradient of pressure seems to bring down the effect of radial velocity which increases the effects of azimuthal velocity and consequently reversal can occur both in radial and the azimuthal directions. It should indeed be an interesting analysis when one takes into account viscous fluids as viscosity also transports angular momentum.

In the previous section, we touched upon the features of the stability of perturbations of self-similar solutions. It is interesting to note that the addition of the Coriolis terms in the fluid equations helps in enlarging the parameter space for the self-similar solutions. Also, it has been noticed that (Mukhopadhyay and Prasanna, 2003) if one has a direct coupling of the two angular velocities, one of the central star and the other of the disk say $\omega = a\Omega$, (a should not be confused with the Kerr parameter) it is first noted that for the reality of the solutions one needs to have the inequality $a > -1/4$, thus putting a restriction on the contrarotating disk. Further, it has also been seen that the rotational effect seems to stabilise the solutions for the parameter range $\frac{4}{3} \leq \gamma \leq \frac{5}{3}$, γ being the gas constant and the viscosity parameter α to vary from 0 to 0.3. Another aspect of interest that followed in this analysis is that the normalised Bernouli parameter 'b' as defined by Narayan and Yi (1994) is given by,

$$b \approx [1 - 2n + 3f(1 + 4a)]/[2n + (5f/\epsilon)(1+4a)], \qquad (12.28)$$

f being the ratio of the advected energy to the heat generated. This means as the corotating fluid can have energy transfer only outwards for $f > 1/3$, the counter-rotating flow can have it either way depending upon Ω and ω. For advection-dominated flows $(f = 1)$, b is mostly positive for $a > -0.2$, while it changes sign for corotating flows at $a = 2 - \sqrt{5}$. These features result in the conclusion that if the energy transfer inwards has to be effective the corotating flow has to have lower angular velocity $\Omega < \omega/(2 + \sqrt{5})$ whereas the counter-rotating flow has to have large angular velocity $\Omega > \frac{\omega}{2 - \sqrt{5}}$ rendering the efficiency of accretion from counter-rotating flows to be more effective for energy transfer.

After these brief discussions on accretion disk dynamics without and with magnetic fields, we take up discussions on other important and formal aspects of relativistic astrophysics dealing with Gravitational lensing, Gravitational radiation and Cosmology.

Chapter 13

Gravitational Lensing

One of the most interesting experiments that all of us as children enjoyed must have been burning a paper by holding a lens in sunlight. As one learned about the light waves and the optical features of lens, it was clear how the light rays passing through a convex lens bend and concentrate towards the focus giving a bright spot of illumination. Even in the Newtonian description of light as corpuscles and gravity as a force between masses, light deflection by gravity was thought about as worked out by Cavendish (1784) and Soldner (1803). This feature of bending of light rays was considered by Einstein (1911), purely on the basis of the equivalence principle. However, in all these approaches one gets only half the value for the deflection of what is noted observationally. It was only after the complete derivation of general relativity in 1915, with the first exact solution of the field equations (Schwarzschild, 1916) one had the exact value for deflection as was observationally confirmed by Eddington (1919). This bending of light in a gravitational field was introduced in Chapter 4 while discussing the topic 'gravity and light', which indeed was a consequence of Einstein's notion of mass — energy equivalence, that really set up the need for a curved space formalism of space-time and study of photon trajectories as geodesics of the underlying geometry. Using this logic Einstein obtained the exact amount of bending of the light ray passing across the solar limb, which was verified and the theory established in 1919 as discussed already. With this bending in mind, Einstein had another thought experiment about the appearance of a double star system one behind the other. His logic led him to predict that if a star is aligned perfectly in the line of the observer and a fore ground star, then the one behind should appear as a closed ring since the rays from that star would get bent around the star in front due to the lens effect. If the alignment is not perfect then the background star image would get split and one may see more than one image. Apparently, Chowlson (1924) before Einstein, had mentioned the appearance of a double star but astronomers did not take it seriously. However, finally it was due to Fritz Zwicky's observation (Schnider, 1992) expressed thus "the formation of multiple images of background objects through gravitational light deflection is not only observable but they actually should be observed", in a published paper "Nebulae as gravitational lenses" (Zwicky 1937) that the investigations ensued to study 'Gravitational lensing'. The interest in this topic increased after the discovery of Quasars and particularly after the detection of a double Quasar 0957+561 also known as a twin Quasar.

Perlick (2004) mentions that "In the most general sense gravitational lensing is a collective term for all effects of gravitational field on the propagation of electromagnetic radiation described as rays. From a mathematical point of view, the theory of gravitational lensing is the theory of light-like geodesics in a four-dimensional manifold with Lorentzian metric".

It is useful to briefly describe the 'lensing geometry" which can explain the feature of multiple imaging of background stars by massive foreground matter distribution. This will indeed help understanding the optical features of cosmic objects, particularly of Quasars necessary for cosmology.

Consider the ray geometry as shown in Fig. 13.1 representing the source S, S_1 and S_2 as images and L being the lens (foreground mass). Let angles $AOS_1 = \theta$, $SOS_1 = \alpha$, $AOS = \beta$, and $SLS_1 = \alpha'$ represent the angles subtended by various arcs. The bending angle α' which in Schwarzschild geometry is given by $4GM(r)/r\,c^2$, $M(r)$ being the mass contained in the volume of radius r, is related to the other angles as given by $\alpha'\,D_{ls} = (\theta - \beta)\,D_s$. Defining a reduced deflection angle $\alpha(\theta) = (D_{ls}/D_s)\alpha'$, one finds $\alpha(\theta) + \beta = \theta$, which when expressed vectorially is called the "lens equation". For a point lens, using $\xi = D_l\theta$ one finds

$$\beta(\theta) = \theta - (D_{ls}/D_l D_s)4GM/\theta c^2. \qquad (13.1)$$

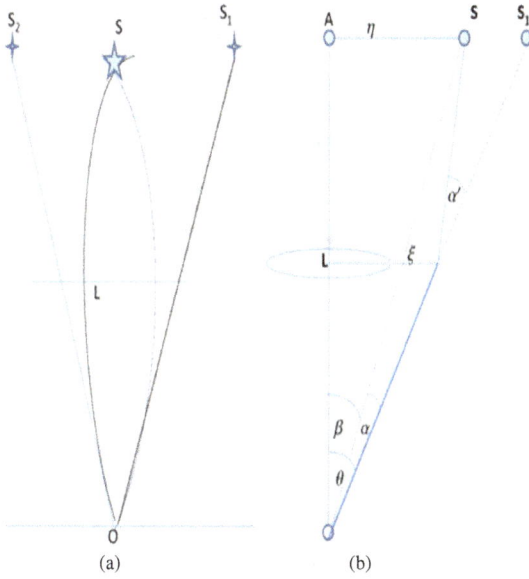

Fig. 13.1.

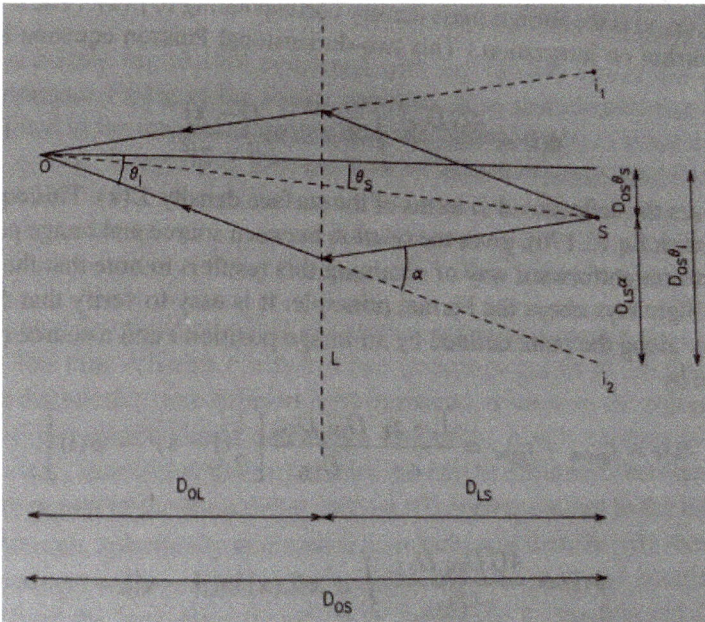

Fig. 13.2.

It is now clear that if the star is located exactly behind the lens then the angle β is zero, which produces a ring-like image with the angular radius

$$\theta_E = \sqrt{\{(D_{ls}/D_l D_s)4GM/c^2\}}. \tag{13.2}$$

called the **Einstein angle** and the ring is called the **Einstein ring**. Using this in the lens equation, one finds the simple equation $\beta(\theta) = \theta - \theta_E^2/\theta$, solving which one gets the solutions $\theta_\pm = (\beta \pm \sqrt{\beta^2 + 4\theta_E^2})/2$, showing that there are two images for the source with one inside and the other outside the ring, appearing on either side of the source. θ_E provides a natural angular size in describing the lensing geometry and it appears that the sources closer than about θ_E to the optic axis are significantly modified whereas the ones outside the ring are not. The angular separation between the two images is given by the difference $\Delta\theta = \theta_+ - \theta_- = \sqrt{\beta^2 + 4\theta_E^2} \geq 2\theta_E$. Looking at the magnification, it is known that due to Liouville's theorem, the number of geodesics in a given cross-sectional area does not change, even though the bundle may get slightly distorted. If the source subtends a solid angle $\Delta\omega_s$ at the observer in the absence of lensing and $\Delta\omega_L$ is the solid angle subtended by the image, as lensing preserves the surface brightness of the source, the flux is $I \cdot \Delta\omega$ with I denoting the specific intensity. For an infinitely small source, the ratio between the solid angles gives the flux amplification (ratio of image area to source area) due to lensing defined as $\mu = d\omega_l/d\omega_s = \theta d\theta/\beta d\beta$. Using the lens equation one can evaluate this to be

$$\mu_\pm = 1/(1 - [\theta_E/\theta_\pm]^4) = [(u^2 + 2)/2u\sqrt{(u^2 + 4)}] \pm 1/2, \tag{13.3}$$

where $u = \beta/\theta_E$ is the angular separation of the source from the lens measured as a function of θ_E called the *impact parameter*. The $\pm$ sign goes with the parity for the outside and inside images which are mirror reflections. The inverse of the magnification factor is represented by

$$1/\mu = \det A = \det(\partial\beta/\partial\theta). \tag{13.4}$$

The sum of the absolute values of the two components gives the total magnitude measurable as given by $[(u^2 + 2)/u\sqrt{(u^2 + 4)}]$. It is clear that as β

tends to zero, magnification diverges with $u = 1$ and $\mu = 1.34$, the values that are taken to typically characterise the efficiency of the lens. The image at a point x of an infinitesimally small source is thus brightened or dimmed by a factor $|\mu|$ as $\mu(x)$ can be positive or negative. The corresponding images are said to have positive or negative parity. Points where the detA vanishes leading μ to diverge are called critical points and at such diverging points the geometrical optics approximation fails (Schnider *et al.*, 1992). The curves on the lens plane along which the Jacobian determinant det $\left[\dfrac{\partial \beta}{\partial \theta} \right]$ vanishes are called *critical curves* and on these curves though the magnification diverges, it does not mean that the image gets infinitely bright as the real sources are extended which always leads to finite magnification. *Caustics* which are basically critical curves in the source plane obtained from the lens mapping are important as they indicate the number of images changing by two if and only if the source crosses a caustic (Schnider *et al.*, 1992). This feature helps in determining the number of images on the position of the source. Blandford & Narayan (1986) have argued that caustics which involve the transition from one topology to another either through a merger or creation of images play an important role in gravitational lensing.

13.1 Observational Features

The first observation of multiple images of cosmic sources was by Walsh *et al.* (1979). When they reported the quasar 0957+56 having two optical components with identical redshifts of the order ≈ 1.045, while the redshift of the lensing galaxy is about 0.355, indicating a distance of about 3.7 billion lys.

Both images have an apparent magnitude of the order ~17, as well as a time lag of about 417 days between the two images. The quasar is found in the constellation Ursa Major, about 10 arc minutes north of NGC 3079. The lensing galaxy YGKOW G1, sometimes called just G1 is a giant elliptical galaxy of type cD which is a part of a cluster that also contributes in the lensing (Wikipedia) and is slightly off the center.

Another example where the lensing produces an Einstein ring which as was mentioned earlier, results only when the source and the lens are in

First find of gravitational lensing, double Quasar 0957+56, also known as 'twin quasar which is located about 14 billion lys from Earth. The two components are only about 5arcsecs away from each other and their spectra look virtually identical. The ratio of magnitudes of these taken at IR wavelengths is very similar to the ratio at visible and UV wavelengths too.

Image: ESA/Hubble & NASA, 2014

Fig. 13.3.

Fig. 13.4.

perfect alignment is depicted in Fig. (13.4). The first such lensing was discovered by Hewitt *et al.* (HTSBL188) who observed the radio source MG1131+0456, using the very large array. They have observed an extremely unusual structure in a radio map with sub-arc second resolution where the object appears as an elliptical ring of emission accompanied by a pair of more compact sources with continuum emission without any emission lines.

Another set of objects that have drawn good attention are the LRGs. The figure (Fig 13.5) shows the picture of a luminous red galaxy (LRG) that has been gravitationally distorted by the light from a much more distant blue galaxy. More typically, such light bending results in two discernible images of the distant galaxy, but here the lens alignment is so precise that the background galaxy is distorted into a horse shoe- a nearly complete ring. Although this lensing galaxy LRG 3-757, was discovered in 2007 in data from Sloan Digital Sky Survey, the image shown above is a follow-up observation taken with the Hubble space telescope's wide field camera 3. Strong gravitational lenses like this are more than oddities where the fore ground galaxy is extremely massive, almost a hundred times that of this galaxy and is notable as it belongs to a rare class of galaxies called luminous red galaxies which have extremely luminous IR emission. This lens is also popularly called as Cosmic Horse shoe (Wikimedia commons).

Thanks to the space telescope Planck-AllSky survey, to analyse gravitationally lensed Extreme Star Burst project, which searched almost 10,000 deg of the sky and found about twenty HyLIRG (Hyperluminous IR galaxies) one of them being PJ0116-24 (Fig. 13.5) is the most recent addition (May 2024) to the group of rarest and most extreme starbursts and found only in the distant universe ($z \geq 1$) and have intrinsic IR luminosities of the order $2.6 \ 10^{14} \ L_{sun}$ and has a red shift $z = 2.125$. It has been found (Liu *et al.*, 24) that the object is highly supported rotationally with a richer gaseous sub-structure than other known galaxies of the same type and also found to be intrinsically massive ($M_{baryon} \simeq 10^{11.3} \ M_{sun}$).

As most of the images in this category are IR images, it is important to see what the most powerful IR space telescope JWebb (launched in Dec 2021) has provided the astronomers with. Below are two pictures, from Webb (Fig. 13.7) and from the Hubble (2017) (Fig. 13.6), of the gravitational lens caused by SDSS J1226+2149, a galaxy cluster located roughly 6.3 billion light years from earth in the constellation ComaBerenices. The lens has amplified the light from the more distant Cosmic Seahorse galaxy. Closer galaxies appear bright and blue-white in the image, whereas the distant ones appear dimmer and reddish due to Cosmological red shift. Comparison with the Hubble picture helps in understanding galactic

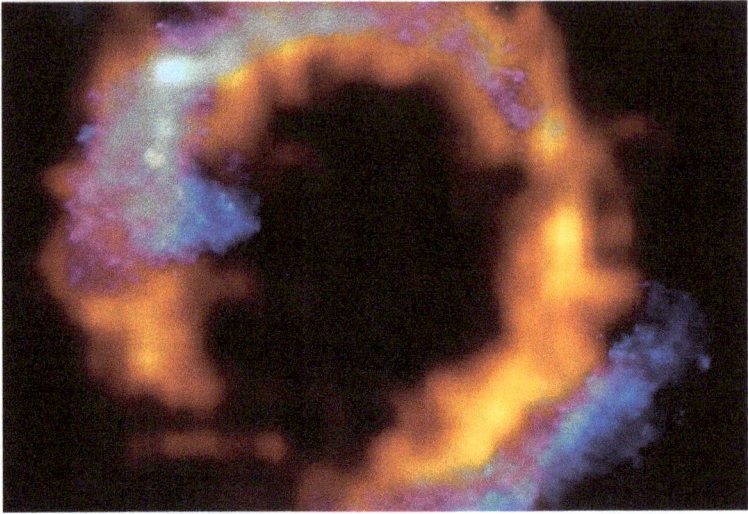

Fig. 13.5.   The distant galaxy PJ0116-24, a Hyper Luminous Infrared Galaxy (HyLIRG). Cold gas is seen here in blue; warm gas is shown in red. Credit: ALMA (ESO/NAOJ/ NRAO)/ESO/D. Liu *et al*.

Fig. 13.6.   Hubble telescope. *Image Credit*: ESA/Webb, NASA & CSA J Rigby.

Fig. 13.7. JWebb telescope.

evolution, particularly of stellar populations from afar (universetoday. com/166788/).

Apart from the above it appears that there are two distinct classes of lensing. (1) Microlensing and (2) weak lensing. Microlensing is a transient phenomenon that occurs when the line of sight sweeps through a dense field of stars that makes the distant compact object (source) appear to have a light curve. Consequently, there appears variation in the flux of the lensed sources depending upon the relative variations in position and velocities of the source. (Schnider *et al.*, 1992) With the possibility of this phenomenon, Paczynski (1986) has pointed out that one could use it to determine whether the Cosmic 'dark matter' appear as discrete lumps by comparing the variability of light curves with expected light curves. This idea led astronomers to look for MACHOS (massive compact halo objects) which was not supported by observations. However, this idea lead to the discovery of extra solar planets which in itself has been a profitable endeavour in astronomy.

Weak lensing deals with effects of light deflection that can be measured only statistically as it arises due to inhomogeneities along the line of sight but does not seem to have yielded any significant results of cosmological interest. Though it is a useful tool to successfully reconstruct the surface mass distribution of clusters, the statistically incoherent weak lens induced change of apparent brightness could affect standard candle-like supernovae that might cause difficulties in the accurate determination of cosmological parameters.

Chapter 14

Gravitational Waves

14.1 Introduction

Since the time the early humans looked up to see the cosmos the only window that was available for understanding the universe has been the electromagnetic window, radiation of different frequencies, optical followed by radio and subsequently IR, UV, X-ray and the γ-rays. However, luckily in this decade since 2015 another window opened for us, the window of gravitational waves. Just as electromagnetic waves are disturbances in the electromagnetic fields, gravitational waves are disturbances in the gravitational fields evolving as perturbations of the associated space-time of massive gravitating sources. The generation and propagation of electromagnetic waves are governed by Maxwell's equations which are linear in character, whereas gravitational waves are governed by Einstein's equations of general relativity which describes the gravitational field as the curvature of the space time associated with mass distributions, and are nonlinear (due to self interaction) in character. As Eddington (1922) says, the problem of the propagation of disturbances of the gravitational field was investigated by Einstein (1916, 1918) who surmised that the disturbances propagate at the speed of light. Weyl had classified plane gravitational waves into three different types *viz.*, longitudinal-longitudinal (LL), longitudinal-transverse (LT) and transverse-transverse (TT), Eddington had shown that while the speed of the first two types was uncertain and therefore suspected their existence, the TT waves propagated with light velocity in all coordinate systems. It is important to note that Einstein's objection for the first two types was also due to the fact that they did not carry energy. Further the Newtonian theory of gravity cannot predict the possibility of gravitational waves, as the governing differential equations are elliptic in nature while Einstein's equations that govern the general theory of relativity are hyperbolic. As superposition of solutions to nonlinear differential equations is not valid unlike the case of electromagnetic wave solutions of Maxwell's equations, Einstein suggested to look for a description in terms of linearized set of gravitational field equations wherein the gravitational field itself is described as a perturbation of flat space-time in the form

$$g_{ij} = \eta_{ij} + h_{ij}, \tag{14.1}$$

η_{ij} being the Minkowski metric of flat space-time $ds^2 = \eta_{ij}dx^i dx^j$. This assumption is quite justified because of the fact that the wave zone is far away from the source and thus the field of the source may be assumed to be weak. As Weber (1966) points out, the problem Einstein considered was that of a rod spinning about one of its axes which gave the radiated power to be about

$$P = 32GI^2\omega^6/5c^3 \approx 1.73 \times 10^{-59} \ I^2\omega^6 \ \text{ergs/sec},$$

where I is the moment of inertia of the rod about the spin axis and ω the angular velocity.

Considering the field outside the matter distribution $R_{ij} = 0$, and using (14.1) one gets the set of equation up to the first order terms in h

$$D_n(h_{ij}) + h^k_{k,ij} - h^k_{i,jk} - h^k_{j,ik} = 0, \tag{14.2}$$

where D_n Stands for the D'Almbertian operator $\left(\frac{\partial^2}{\partial t^2} - \frac{\partial^2}{\partial x^2} - \frac{\partial^2}{\partial y^2} - \frac{\partial^2}{\partial z^2}\right)$.

As solutions to this equation cannot be unique due to general covariance one needs to choose a particular gauge. One generally assumes the so called Lorentz or deDonder gauge as given by the condition $g^{ij}\Gamma^k_{ij} = 0$, which in terms of h yields the equations

$$\bar{h}^j_{i,j} = 0, \quad \bar{h}^j_i h^j_i - \frac{1}{2}\delta^j_i h^k_k. \tag{14.3}$$

This choice of the gauge, reduces the equations to a simpler form $D_n(h_{ij}) = 0$. As these differential equations are of second order, linear and hyperbolic one can write the general solution as a superposition of set of monochromatic plane waves as described by

$$h_{ij} = A_{ij}e^{ik_nx^n} + A^*_{ij}e^{-ik_nx^n}. \tag{14.4}$$

Here, A and A^* are the complex amplitudes, k^i the wave vector satisfying the orthogonality relation $\eta_{ij}k^ik^j = 0$, representing the characteristics of the system which are indeed the null hypersurfaces of the background flat space -time. This basically says that the disturbances (perturbations of the flat space time, the gravitational waves) do propagate along the null geodesics of the background space. It is important to check on the constraints put in by the gauge condition introduced.

As there are ten components of the complex amplitude tensor A_{ij} the gauge condition yields four constraints as given by the relation $A_{ij}k^j = \frac{1}{2}A^j_jk_i$. However, as still some coordinate freedom is left one can choose a globally defined time like vector u^i such that $A_{ij}u^j = 0$, along with making A^i_i, the trace of the tensor A^j_i equal to zero. Thus the ten components of the tensor A_{ij} satisfy eight conditions: $A_{ij}k^j = 0$, $A_{ij}u^j = 0$, and $A^j_i = 0$, rendering only two components to be independent. As this comes purely from the choice of gauge, it is called the 'Transverse–Traceless gauge'. These conditions expressed in terms of the gravitational potential h_{ij} yield

$$h_{i0} = 0, \quad h^j_{a,j} = 0, \quad h^i_i = 0 \quad (a = 1-3). \tag{14.5}$$

The two independent degrees of freedom indicate that there are only two possible polarization states for the waves which are linked to the fact that the plane gravitational waves are with helicity ±2. Those interested in more details may refer to Weinsberg (1972) and Prasanna (2017).

14.2 Relative Acceleration

In Chapter 3, while discussing the gedanken experiment of freely falling lift the notion of relative acceleration between freely falling particles in a

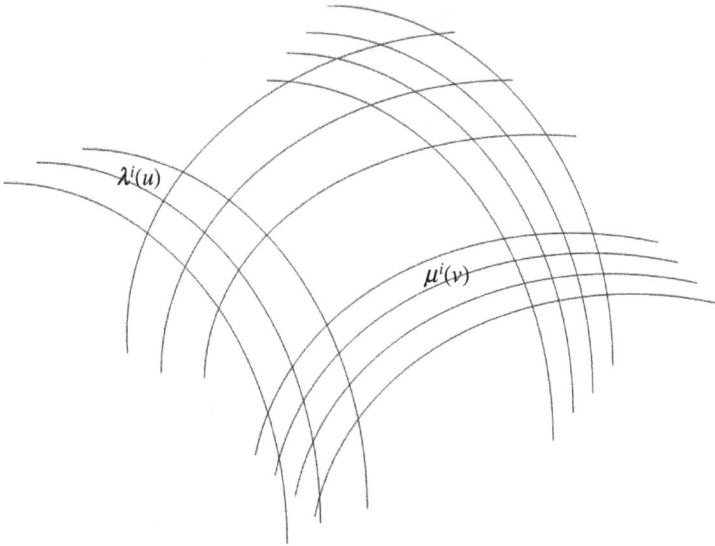

Fig. 14.1.

gravitational field was mentioned with the formula as derived in Newtonian physics. Considering the situation in general relativity one needs to lookout for particles following geodesics and their interaction which can be derived through the equations of geodesic deviation. Considering a congruence of geodesics, one can assume a two surface S defined as $x^i = x^i(u,v)$, with u denoting the proper time along the geodesics and v labelling distinct geodesics as shown in the diagram. Then one can define two distinct vector fields as given by $U^i = dx^i/du$ and $V^i = dx^i/dv$ that are tangents to the families of curves $\lambda^i(u)$ and $\mu^i(v)$, respectively.

It is simple to verify that the covariant derivative of U^i along μ, $U^i_{;j}V^j$ denoted as $\delta U^i/\delta v$ and of V^i along λ, $V^i_{;j}U^j$ denoted by $\delta V^i/\delta u$ are equal in a Riemannian space where the connection is symmetric (torsion free).

Choosing any two particles and their worldlines (geodesics λ_1, λ_2) as was done in the Newtonian case, one can identify U^i with the tangent to the geodesic and the connecting vector η^i with V^i one will have $\delta \eta^i/\delta s = \delta U^i/\delta v$.

This will give for the second derivative $\dfrac{\delta^2 \eta^i}{\delta s^2} = \dfrac{\delta}{\delta s}\left(\dfrac{\delta U^i}{\delta v}\right) = \dfrac{\delta^2 U^i}{\delta s \delta v}$ (14.6)

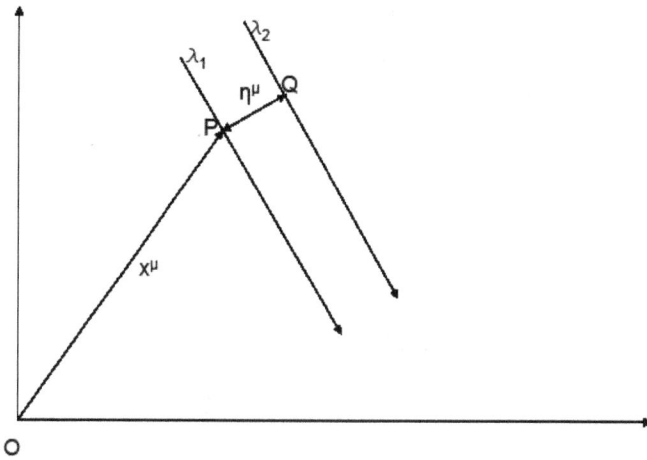

Fig. 14.2.

Considering the difference

$$\frac{\delta^2 U^i}{\delta s \delta v} - \frac{\delta^2 U^i}{\delta v \delta s} = U^i_{;k} V^k_{;j} U^j + U^i_{;k;j} V^k U^j - (U^i_{;k} U^k_{;j} V^j + U^i_{;k;j} U^k V^j) \quad (14.7)$$

The first and third terms on the right hand side cancel each other as by definition $V^k_{;j} U^j = U^k_{;j} V^j$. Rearranging the other two terms one will have on the right hand side $(U^i_{;j;k} - U^i_{;k;j}) U^k V^j$, which again from the definition of the Riemann–Christoffel curvature tensor gives $R^i_{ijk} U^l U^k V^j$, finally resulting in the equation

$$\frac{\delta^2 \eta^i}{\delta s^2} = \frac{\delta^2 U^i}{\delta v \delta s} + R^i_{ljk} U^i \eta^j U^k. \quad (14.8)$$

Here again the first term on the right hand side is identically zero because $\delta U^i / \delta s = 0$, as U^i s are geodesics and s is the path parameter. Thus, finally one has the equation of geodesic deviation as given by

$$\delta^2 \eta^i / \delta s^2 + R^i_{ljk} U^l \eta^k U^j = 0. \quad (14.9)$$

This equation will play a very important role in analyzing the gravitational wave signals.

In order to understand the physical meaning of this equation, consider a local Lorentz frame tetrad λ^i_j, where the time-like component (λ^i_0) is along the tangent to the geodesic and the space like triad $\lambda^i_a (a = 1.3)$ is propagated parallely along the geodesic such that the physical components of any vector V are expressed as $V^i = \lambda^i_a X^a$. Decomposing the vector η^i and the tensor R^i_{ljk} along the tetrad, equation (14.9) takes the form

$$\lambda^i_a(d^2 X^a / ds^2) + \lambda^i_a R^a_{bcd} \lambda^b_l \lambda^c_j \lambda^d_k U^l U^j \lambda^k_m X^m = 0. \tag{14.10}$$

Using the fact that the tetrad is orthonormal $\lambda^d_i \lambda^i_m = \delta^d_m$ one gets finally

$$\lambda^i_a[(d^2 X^a / ds^2) + R^a_{ljd} U^l U^j X^d] = 0. \tag{14.11}$$

As this is true for all choice of the tetrad λ, this may be written as

$$d^2 X^a / ds^2 + K^a_d X^d = 0 \quad K^a_d = R^a_{ljd} U^l U^j \tag{14.12}$$

which is in the same form as in the case of Newtonian mechanics. This further emphasizes the interpretation of the *curvature being the manifestation of the acceleration in a gravitational field.*

As was mentioned in Section 14.1, the main attempts to analyse the gravitational waves, comes from its description as a radiative solution in a weak field approximation of Einstein's equations for empty space as expressed through the equations (14.3 and 4, for plane waves) along with the gauge conditions (14.5 and 6). To understand the two states of polarisation for the wave it is instructive to analyse the effect of a passing plane wave on a circular ring of particles, using the equations of geodesic deviation that characterises the curvature through the induced relative acceleration as described above.

Considering a ring of particles as shown in Fig. 14.3(a) and a gravitational plane wave described by the metric

$$ds^2 = dt^2 - (1 - h_{XX})dx^2 - (1 - h_{YY})dy^2 + 2h_{XY}dxdy - dz^2. \tag{14.13}$$

The only non zero components of the curvature tensor for the metric are

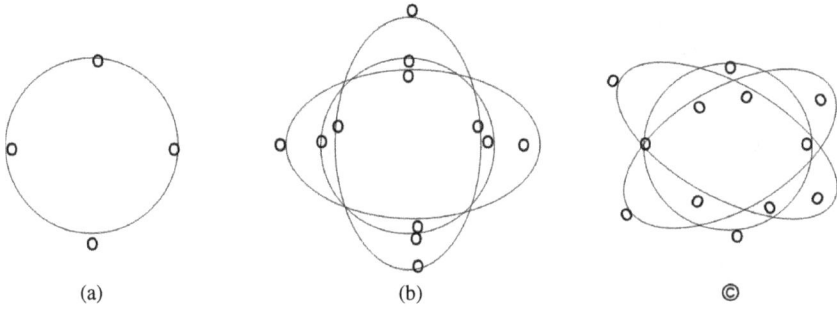

Fig. 14.3. Gravitational wave passing through a ring of particles (a) before (b) wave with + polarisation (c) wave with x polarization. Adopted from Satyaprakash and Schutz (2009).

$$R^x_{0x0} = -\frac{1}{2}h^{TT}_{XX,00}, \quad R^y_{0y0} = -\frac{1}{2}h^{TT}_{YY,00} \quad R^x_{0y0} = R^y_{0x0} = -\frac{1}{2}h^{TT}_{XY,00}.$$

Choosing a commoving frame $U^i = (1, 0, 0, 0)$ and the deviation vector $\eta^i = (0, \varepsilon, 0\ 0)$ the geodesic deviation equation yields,

$$\partial^2\eta^x/\partial t^2 = \frac{1}{2}h^{TT}_{XX,00}\,\varepsilon, \quad \partial^2\eta^y/\partial t^2 = \frac{1}{2}h^{TT}_{XY,00}\,\varepsilon.$$

On the other hand if the deviation vector $\eta^i = (0, 0, \varepsilon\ 0)$ then the equations are

$$\partial^2\eta^x/\partial t^2 = \frac{1}{2}h^{TT}_{XY,00}\,\varepsilon, \quad \partial^2\eta^y/\partial t^2 = -\frac{1}{2}h^{TT}_{XX,00}\,\varepsilon.$$

These equations indicate that the particles in the ring gets accelerated accordingly and the resulting structures are depicted in Figs. 14.3 (b,c).

As the amplitudes h_{XX} and h_{XY} are independent the two sets of oscillations represent two independent polarisation states that are at a relative orientation of 45^0 with respect to each other, one can express the amplitudes as

$$h_{XX} = -h_{YY} = a_+ \text{Sin}\ (\omega(t - z) + \varphi_+),\ h_{XY} = -h_{XY} = a_X \text{Sin}\ (\omega(t - z) + \varphi_X).$$

If in a special case $a_+ = a_X$, and $\varphi_X = \varphi_+ \pm \pi/2$, it corresponds to the case of circular polarisation.

14.3 Energy and Momentum Carried by Plane Waves

As shown in several standard texts (LL, MTW, SW) the gravitational field energy cannot be localised which makes it impossible to differentiate between the source energy and the field energy from the total energy momentum that appears in the field equations. Luckily in the context of the linearised equations (with perturbations on flat geometry) one can define a pseudo tensor t_i^j that characterises the energy of plane gravitational waves. From the field equations $G_i^j = T_i^j$, using the conservation law $T_{i;j}^j = 0$,

$$\frac{1}{\sqrt{-g}}\left[\partial(T_i^j \sqrt{-g})/\partial x^j\right] - 1/2\,(\partial g_{jk}/\partial x^i)T^{jk} = 0 \qquad (14.14)$$

and expressing it in the form $\frac{\partial}{\partial x^j}[-g(T_i^j + t_i^j)] = 0$, one can see that the total energy momentum has been separated into the source part (T_i^j) and the field part (t_i^j) that can be defined through a super potential ψ^{ikl}, Landau-Lifshitz (1951) given by the field equation and can be written as (Anderson, 1972)

$$\psi^{ikl} = \sqrt{-g}\,\delta_p^i\left[g^{kp}g^{lm} - g^{km}g^{lp}\right]_{,m} \qquad (14.15)$$

It may be noted that with this definition, the total energy-momentum $[-g(T_i^j + t_i^j)]$ can be called the 'effective energy-momentum' of the space-time governed by the chosen metric that satisfies the divergence free relation $(\partial/\partial x^j)(T_i^j + t_i^j) = 0$, which can be used to get the effective energy through a volume integral. In the case of linearised gravity with perturbations of the flat metric, the short wave approximation defined by $\lambda/\mathcal{R} \ll 1$ and $a \ll 1$, with λ being the wave length and $\mathcal{R}$ the scale length, the Ricci tensor can be written as $R_{ij} = R_{ij}^B$ + terms of $\vartheta(h)$ + terms of $\vartheta(h^2)$ + error of $\vartheta(a^3/\lambda^2)$ (Misner *et al.*, 1973). As the background metric is flat, one thus finds for the averaged stress energy pseudo tensor (averaged over several wavelengths) the expression

$$t_{ij} = 1/8\pi\{< R_{ij}(h^2) > -1/2 g_{ij}^B < R(h^2) >\} \tag{14.16}$$

which in the TT gauge yields $<t_{ij}> = 1/32\pi <h^{kl}>_{,i} h_{kl,j}$. (14.17)

When the averages are taken over one period of oscillation in time and spatial regions of the size of a wavelength in all directions this expression is also called the Issacoson stress-energy tensor for gravitational waves. For the current metric under consideration (14.13) the non zero components of this stress tensor are given by

$$t_{00} = t_{zz} = -t_{0z} = \frac{1}{32\pi} w^2 (mod\, a_+^2 + mod\, a_x^2). \tag{14.18}$$

It is thus seen that the plane gravitational waves on a flat background move along null geodesics ($\eta_{ij}k^i k^j = 0$) possessing two independent states of polarisation carrying energy-momentum proportional to square of their amplitudes.

14.4 Generation of Gravitational Waves

The waves that are propagated and observed at infinity (far zone) must have originated at the source due to mass motion at the source distribution (near zone) and when the mass motion is slow ($v \ll c, v^2/c^2 \sim \frac{2m}{r}$) the weak field approximation is best suited to analyse the dynamics. Accordingly the field equations may be written in the harmonic gauge ($h_i^j - \frac{1}{2}h\delta_i^j)_{,j} = 0$, as

$$\Box \bar{h}_{ij} = -2\kappa T_{ij}, \quad \tau^j{}_{i,j} = 0. \tag{14.19a}$$

whose solution may be written in terms of retarded Green's function which after integration with respect to t gives

$$\bar{h}_{ij} = 4\int \left\{ \left[\frac{T_{ij}(x', t - |x - x'|)}{|x - x'|} \right] \right\} d^3 x'. \tag{14.19b}$$

From the conservation laws one can write

$$(a) \; \tau^{ab}{}_{,b} + \tau^{a0}{}_{,0} = 0$$

$$(b) \; \tau^{0b}{}_{,b} + \tau^{00}{}_{,0} = 0. \tag{14.20}$$

Taking the first moment of (14.20a) and the second moment of (14.20b), one gets

$$\frac{\partial}{\partial t}\int \tau^{a0}x^c dv = -\int \frac{\partial \tau^{ab}}{\partial x^b}x^c \, dv = -\int \frac{\partial(\tau^{ab}x^c)}{\partial x^b}dv + \int \delta^c_b \, \tau^{ab} \, dv$$

$$\Rightarrow \frac{1}{2}\frac{\partial}{\partial t}\int (\tau^{a0}x^c + \tau^{c0}x^a)dv = \int \tau^{ac} \, dv, \tag{14.21}$$

and

$$\frac{\partial}{\partial t}\int \tau^{00}x^a x^b dv = \int \frac{\partial}{\partial x^c}(\tau^{0c}x^a x^b)dv + \int \tau^{0c}\frac{\partial}{\partial x^c}(x^a x^b)dv. \tag{14.22}$$

which implies

$$\frac{\partial}{\partial t}\int \tau^{00}x^a x^b dv = \int (\tau^{0b}x^a + \tau^{0a}x^b)dv \tag{14.23}$$

These equations together give

$$\int \tau^{ab}dv = \frac{1}{2}\frac{\partial^2}{\partial t^2}\int \tau^{00}x^a x^b dv \tag{14.24}$$

As the discussion relates only to slow motion approximation, the only contribution to the stress energy tensor is from the mass density ρ, the final equation is

$$\int \tau^{ab}dv = \frac{1}{2}\frac{\partial^2}{\partial t^2}\int \rho(r,t)x^a x^b dv = \frac{1}{2}I^{ab} \tag{14.25}$$

I^{ab} being the second moment of the mass distribution at the source related to the moment of inertia tensor (Misner *et al.*, 1973)I^{ab}

$$I^{ab}\int \rho(r^2\delta^{ab} - x^a x^b)dv = -(I^{ab} - \delta^{ab}I) \tag{14.26}$$

and to the 'Quadrupole moment' (Landau, 1932)

$$Q^{ab} = \int \rho(x,t)[3x^a x^b - r^2 \delta^{ab}]dv = (3I^{ab} - \delta^{ab}I) \qquad (14.27)$$

(I being the trace of the tensor I^{ab}).

With these one can now write the approximate solution for the field equations (14.19a) to be

$$\bar{h}_{ij} = (-2\Omega^2/r)I_{ij}\exp[i\Omega(r-t)] \qquad (14.28)$$

Ω being the frequency. This solution is known as the **Quadrupole formula**.

14.5 Detection of Gravitational Waves

Though it was in 2016 that the first detection of gravitational wave was realised by the LIGO collaboration, it was already in the 60s that Joe Weber had mooted the idea of resonant mass detector and even had developed the prototype in 1966 — a large aluminium cylinder held at room temperature in a vacuum chamber and isolated from possible vibrations. The idea was to see if the passing gravitational wave would induce mechanical vibrations at the resonant frequencies in the bar, due to tidal acceleration between particles of the cylinder which could be detected through piezo electric crystals. The electric current so produced may be measured and if it appears in two such bars positioned about 1,000 kms apart coincidentally would suggest the source of the signals must be due to some event in the cosmos that could generate gravitational waves. In 1969, he announced the possible detection but unfortunately no other group working on similar set up confirmed the claim. Further theoretical calculations about the source at the distance claimed, showed an enormous amount of energy to have been emitted by the source which was found to be unreasonable. However attempts are still on by few other different groups to increase the sensitivity by cooling the cylinder and several other means and yet no significant results have come from bar detectors.

Beam mode detectors use laser beams in a Michelson type interferometer where the beam is split by a beam splitter and the two split beams travel along sufficiently long cavities and get reflected by mirrors at the ends. The cavities are placed orthogonal to each other so that the reflected beams can interfere at the eye piece due to the variation of their path lengths caused by the passing wave affecting the system (along and the perpendicular directions). A photodiode is used to monitor the changes in the interference patterns. (The calibration of the interferometer assures that there is no light reaching the diode if the cavities are of same length.) The passing of a gravitational wave induces a path length change due to tidal accelerations of the particles of the mirrors. The change $DL(t)$ is directly proportional to the output of the photodiode. As Thorne (1999) explains the two modes of polarisation of a gravitational wave the $+$ and the x have their respective gravitational wave field $h+$ and h_x respectively which oscillate in time and propagates with velocity c producing tidal force on any set of particles as shown earlier. When a gravitational wave with frequency higher than the pendulum frequency of about 1 Hz, passes such a configuration it pushes the mirror masses back and forth relative to each other there by affecting the arm-length differenceΔL , which may be expressed as

$$\Delta L/L = F_+ h_+(t) + F_X L_X(t) \equiv h(t). \tag{14.29}$$

F_+ and F_X usually of order unity that depend upon the direction to the source and the orientation of the detector in a quadrupolar manner (Thorne 1987). The combination of Fs in equation (14.29) is referred to as the strain of the gravitational wave and the time evolutions of hs are sometimes called as wave forms. In the early 70s, the progress in this mode of detection was not appreciative till 1974 and the discovery of the binary pulsar PSR 1913 + 16. Hulse and Taylor discovered this first binary pulsar from a routine search from the Arceibo observatory and was identified to be a pair of neutron stars of almost equal mass ($M_p = 1.39 \pm 0.15 M_\odot$ and $M_c = 1.44 \pm 0.15 M_\odot$) moving in a highly elliptic orbit with eccentricity (e = 0.617155 $\pm$ 0.000007) and having a projected semi major axis a Sin $i\sim 7 \times 10^{10}$ cms. (Hulse and Taylor 1976). Understanding the significance of this binary

Table 14.1.   Measured Orbital Parameters for B 1913 + 16 System.

Fitted parameter	Value
$a_p \sin i$(s)	2.3417725 (8)
ω_0	292.54487 (8)
e	0.6171338
$\dot{\omega}$(deg/yr)	4.226595 (5)
T_0	52144.90097844 (5)
γ(s)	0.0042919 (8)
P_b	0.322997448930 (4)
$\dot{P}_b(10^{-12}s/s)$	–2.4184 (9)

pulsar, continuous observations were carried out and consequently after 22 years (1981–2003) yielded an excellent set of orbital parameters which were summarised by Weisberg and Taylor (2005) as presented in the Table 14.1.

It has been suggested that while the first five parameters are derivable just from standard orbital mechanics, the mean rate of advance of periastron $< \dot{\omega}_i >$, gravitational redshift and time dialation factor γ and the orbital period derivative $\dot{P}_b$, do require general relativity. In this sense the binary pulsar is said to be an excellent 'laboratory in the sky' for testing relativistic theories of gravitation.

As the binary period governs the motion of the two comoponents in their orbit the most important way of their losing the orbital energy is through emission of gravitational radiation. Wagoner (1975) and Esposito and Harrison (1975) have shown that as derived by Peters and Mathews (1963) this decrease in orbital period is expressed as

$$\dot{P}_b = \frac{192\pi G^{\frac{5}{3}}}{5c^5} \left(\frac{P_b}{2\pi} \right)^{-5/3} (1-e^2)^{-7/2} \left(1 + \frac{73}{24}e^2 + \frac{37}{96}e^4 \right)$$
$$[m_p m_c / (m_p + m_c)^{-1/3}]. \tag{14.30}$$

The relativistic variables $< \dot{\omega} >$ and γ both depend upon the masses of the components and expressed as

$$<\dot{\omega}> = 3G^{2/3}c^{-2}\left(\frac{P_b}{2\pi}\right)^{-5/3}(1-e^2)^{-1}(m_p+m_c)^{2/3} \qquad (14.31)$$

$$\gamma = G^{2/3}c^{-2}e\left(\frac{P_b}{2\pi}\right)^{1/3}m_c(m_p+em_c)(m_p+m_c)^{-4/3} \qquad (14.32)$$

and are measurable. Using the observed values in the above one can solve for the masses and for the binary pulsar the masses turn out to be

$$m_p = (1.4408 \pm 0.0003)M_\odot, \quad m_c = (1.3878 \pm 0.0003)M_\odot, \qquad (14.33)$$

Using these values in (14.5.2) one gets for the orbital period decay $(\dot{P_b})_{GR} = (-2.40247 \pm 0.0002) \times 10^{-12}$ sec/sec. In the early nineties the study of supernovae red shifts indicated that the universal expansion is increased by the acceleration of the galaxies many fold more than the Hubble recession. While on the one hand serious studies were taken up to unravel this feature, it also indicated that this acceleration could influence the orbital periods of pulsars. Studies by Damour and Taylor (1991) concerning this effect along with the motion of the Sun indicated both theoretically and observationally yielded the result $(\dot{P_b}/P_b)^{obs} = 86.79 \pm 0.19(gal) \pm 0.65(obs) \times 10^{-18}/sec$ and the corresponding theoretical estimate was

$$(\dot{P_b}/P_b)^{GR} = -86.0923 \pm 0.0025(gal) \times 10^{-18}/sec.$$

The ratio of the observed to the theoretical estimate of the periods thus turns out to be $(\dot{P_b})^{obs}/(\dot{P_b})^{GR} = 1.0081 \pm 0.0022(gal) \pm 0.0076(obs)$.

This orbital decay in period coming from the emission of gravitational radiation damping should cause a shift in the epoch of peri-astron and as the Fig. 14.4 shows the agreement between the predicted curve over the years match perfectly with the observed points for different years during the period 1975 to 2022. As a result of their loss of orbital energy to gravitational waves the neutron stars of the binary are gradually spiralling inward at the rate almost exactly as predicted by general relativity which is considered as an indirect confirmation of the existence of gravitational waves.

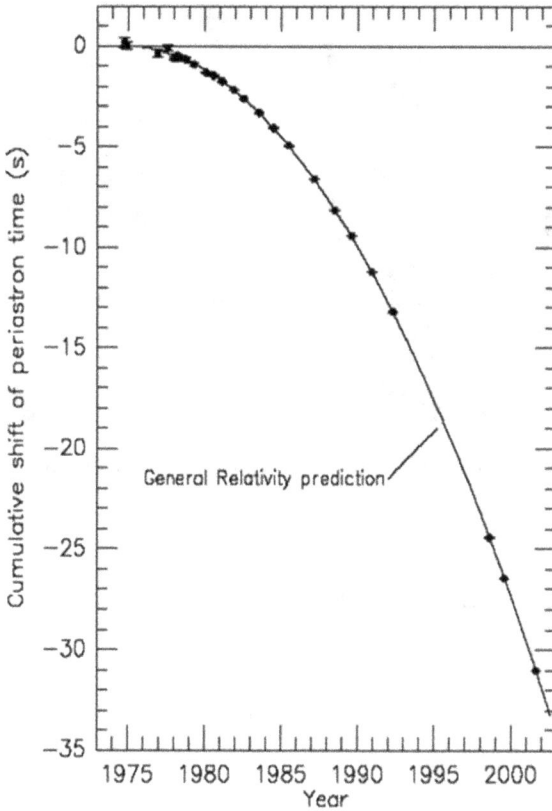

Fig. 14.4.

14.6 Interferometric Detectors

As Thorne (1997) mentions, a laser interferometric gravitational wave detector consists of four masses hung from vibration free supports along with a high sensitive optical detector for detecting the signals through monitoring the lengths (separation) between the mirrors which changes with the passing of a gravitational wave.

A typical wave form arising out of inspiralling compact binary system appears as shown in Fig. 14.6. which has been computed using Newtonian gravity for the orbit evolution and the quadrupole-moment approximation for wave generation (Abromovici *et al.*, 1992).

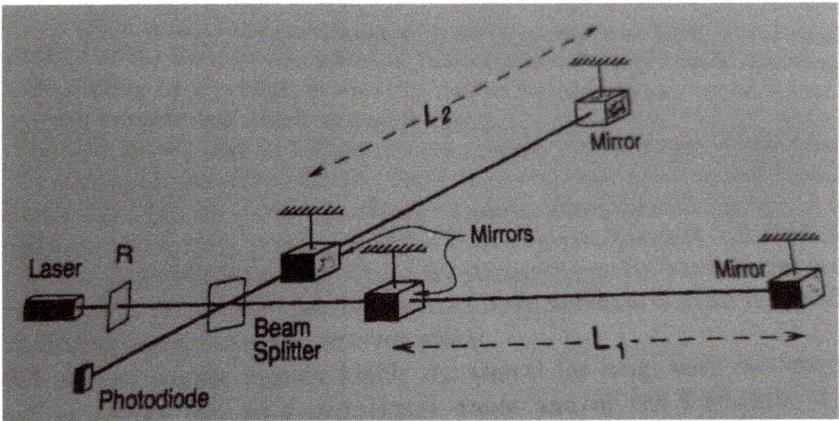

Fig. 14.5.   Block diagram for the beam type detectors.

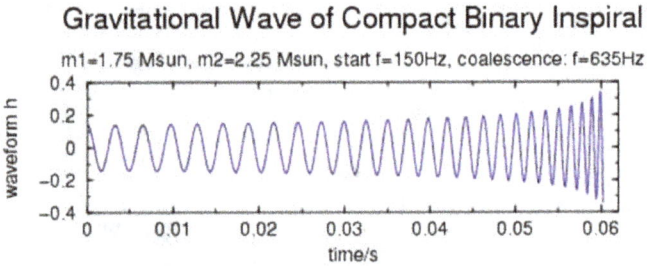

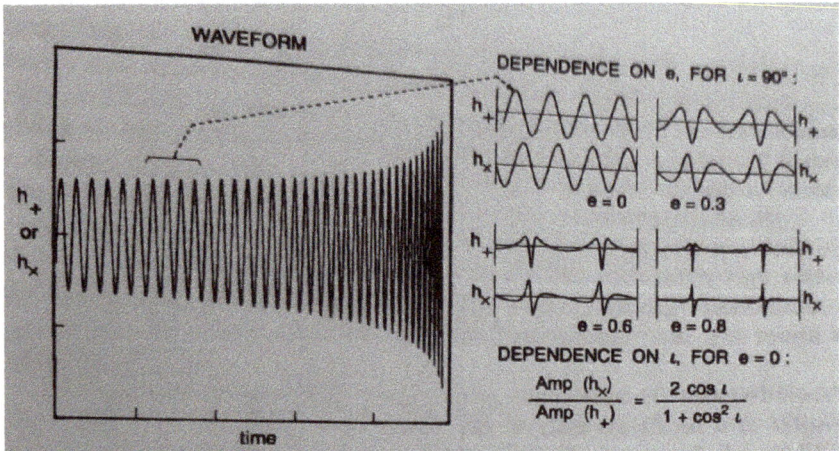

Fig. 14.6.   Typical wave form.

As the inspiralling components of the binary get closer the amplitude increases with an upward sweeping frequency (normally referred to as '*chirp*' of the wave form whose amplitude ratio for the two polarisations is given by $amph_+/amph_x = 2 \cos i / 1 + \cos^2 i$,

Where i is the inclination angle of the orbit to observer's line of sight. The wave's harmonic content is determined by the orbital eccentricity. As an example, if the orbit considered is circular then the rate at which the frequency sweeps or chirps df/dt is determined solely by the binary's chirp mass expressed as

$$M_c = [(M_1 M_2)^3/(M_1 + M_2)]^{1/5}.$$

The ratios of the amplitudes of the two polarisations is determined by the inclination i of the orbit to our line of sight whereas the waves harmonic content (shapes of the individual waves) is determined by the orbital eccentricity. The amplitudes of the wave forms also are determined by the chirp mass in this Newtonian/quadrupole approximation. Thus by measuring the two amplitudes, the frequency sweep and the harmonic content of the inspiral waves one can measure the source's distance, inclination, chirp mass and the eccentricity. Along with these the general relativistic effects add further information through the wave form modulation which comes from the rate of frequency sweep df/dt, depending upon the binary's dimensionless ratio $\eta = \mu/M$, μ being the reduced mass equal to $M_1 M_2/(M_1 + M_2)$ and the spins of the two bodies. Of these two important effects to be noted are,

(i) the back scattering of waves due to the curvature of the binary space-time (Vishvesawra, 1970) which produces tails that act back on the binary and modifies the inspiral rate which can be measured and

(ii) the Lense-Thirring drag arising from the inclinations of the spin axes of the components with respect to the binary's orbital plane, that causes the precession of orbit which also can modulate the wave forms. (Cutler *et al.*, 1993).

The methodology for incorporating these GR effects is through a process called 'matched filter' used while detecting the signals. In this method, the incoming signals are cross correlated to already prepared templates with several different inputs (combination of parameters) and the best matched one is considered for picking the wave form. In order to get all the information in the inspiral wave forms, the templates need to be accurate enough for given masses and spins to a fraction of a cycle during the entire sweep through the observed band. In the case of chirping binaries, post Newtonian theory has been used to model the dynamics of the system to a very high order in v/c. Particularly for signals from binaries of almost equal mass or nearly equal mass and the system being still in the far away zone of coalescence, this method seems effective enough with the wave form taken to be valid till the last stable circular orbit. Even if the components of the binary are tidally disrupted the method is useful as the effects of such disruption are likely to appear at frequencies with sensitivities far beyond the earth based detectors (Satyaprakash and Schutz, 2009).

As Blanchet points out (Blanchet, 2014) the basic problem that one faces in relating the amplitude h_{ab} seen in the wave zone with the source material stress energy tensor is due to the approximations of general relativity. While the post Newtonian methods appear satisfactory in the weak field limit, its inadequacy appears while trying to include the boundary conditions at infinity which affects while calculating the radiation reaction force. While the post Minkowskian approximation appears valid for all over space-time as long as the source is weakly gravitating it would show hurdles while treating the multipole approximation outside the source with respect to the far zone expansion. Further to analyse the final stages of the inspiralling binaries from the data obtained by the ground based detectors one requires very high accuracy templates according to GR, which could be achieved through higher order post-Newtonian wave generation formalism. This was indeed achieved and implemented for the case of detectors LIGO and VIRGO with approximation corresponding to 3PN order (c^{-6}) for neutron star binaries beyond the quadrupole moment (Damour *et al.*, 1998) and several others. However, these calculations though looked satisfactory for the binaries consisting of neutron stars and white dwarfs, for binary black holes, particularly when one of them is massive it is found that techniques of numerical relativity would be necessary as full solution of Einstein's equations are required. As reviewed by

(Centrella *et al.*, 2010) mergers of comparable mass black hole binaries which are expected to be amongst the strongest source of gravitational radiation, the final death spiral consists of three stages called inspiral, merger and ringdown phases. During the first phase the binary orbits get circularised due to the gravitational wave emission and also the components spiral together in quasi circular orbits, as the orbital time scale would be much shorter than the time scale over which the other orbital parameters change, one can treat the components as point particles due to the large separation and consequently apply the orbital dynamics as was in the case of neutron star binaries. (Peters and Mathews, 1963). The wave forms can be calculated using the post Newtonian order $v^2/c^2 \approx GM/Rc^2$, R being the binary separation and one finds that the waveform would have the characteristic of 'chirps'. As the black holes get closer the strong field effects of GR require numerical techniques (three dimensional simulation). At this stage the components are close enough to merge and form a single bigger black hole though distorted, which would finally settle down as a Kerr black hole after radiating away all non axisymmetric modes which is referred to as the ring down phase.

An order of magnitude estimate for the amplitudes of the waves emitted for different phases is as follows. In the *inspiral phase*: the two stars of almost the same mass in a circular orbit of radius R have all their mass in non-spherical motion, such that $(Mv^2) = M(\Omega R)^2 = M^2/R$, with Ω being the orbital angular velocity. The gravitational wave amplitude may then be written as $h_b = 2M^2/rR$. Replacing R by the orbital angular frequency Ω, one finds $h_b = (2/r)M^{5/3}\Omega^{2/3}$. The luminosity of the gravitational wave for such a system turns out to be approximately $L_b \sim 4/5 \ (M/R)^5$. As the orbital radius shrinks the emitted frequency increases towards a chirp, due to radiation emission with chirp time for equal mass binary to be given by

$$t_{chirp} = Mv^2/2L_b \sim (5M/8)/(M/R)^4.$$

As the binary evolves the frequency and the amplitude of the wave grow which influences the binary to evolve more rapidly. However as the inspiral phase ends with the binary reaching *the merger phase* with the distance between the components getting closer to the last stable orbit $(R \sim 6M)$ the frequency reaches the value $f_{LSO} \sim 220\left(\frac{20M_\ast}{M}\right)$Hz, which is referred to as last stable orbit frequency. (Sathyaprakash and Schutz 2009). After this phase the final stage of coalescence gets to the ringdown phase when the emitted gravitational radiation scattered off the finally formed blackhole will have a characteristic wave form called *quasi normal mode* depending upon the size of the blackhole and the incident wave has frequencies above a certain value. This feature was discovered by Vihveswara (1970) while discussing the scattering of gravitational radiation off a blackhole through the discussion of the stability of the blackhole solution of Einstein's equation using perturbative procedure. In fact the observation of these modes (QNMs) is considered as a test for strong field effects of gravity as described by general relativity. (Satyaprakash and Schutz, 2009). For a detailed discussion on this topic of blackhole perturbations, one should refer to Chandrasekhar, 1992).

In the case of unequal mass binaries the coalescing time as measured from the rate of period change $\dot{P}_b = -\frac{192\pi}{5}(2\pi M_c)^{5/3} v$, is $t_{\text{chirp}} = (5M/96\nu)/(M/R)^4$, M being the total mass of the two components. From these various estimates one finds that while binaries with large mass ratios can spend long time in highly relativistic orbits those with almost equal mass are expected to merge after being in this regime for only few orbits. It is estimated that the well known Hulse-Taylor pulsar (discussed above) is expected to merge in just about 300 million years as the orbit is shrinking at the rate ~3.1 *mm/orbit*.

As Satyaprakash and Schutz (2009) point out these perturbed blackholes that exhibit QNMs of vibration emit gravitational radiation of amplitude, frequency and damping time characteristic of the formed blackhole's mass and angular momentum which are the only two features of a Kerr blackhole. The effective amplitude of the waves is of the form $h_{eff} \sim 4\alpha v M/\pi r$. For a pair of 10 M_{sun} blackholes at a distance of 200 Mpc it turns out to be $\left(\frac{v}{0.25}\right)\left(\frac{M}{20M_{\text{sun}}}\right)\left(\frac{200Mpc}{r}\right)10^{-21}$.

As the discussion is basically for relativistic applications, and as the equations of GR are nonlinear partial differential equations it was realised that analytical approaches are not possible in realistic situations and thus one needs to approach numerical techniques for these studies. Hahn and Lindquist (1964) were supposed to be the first to try the simulation approach for studying the dynamics of head on collision of two equal mass blackholes using a two dimensional axisymmetric approach but to their disappointment found that the scheme was not accurate after fifty steps. In spite of a large number of groups studying the numerical simulation for developing codes for the analysis of 3d relativistic hydrodynamics during 1999 to 2006, the first successful simulation was obtained by Pretorius (2005, 2006) which was applied to the data analysis of LIGO-VIRGO observations in 2015 that resulted in the discovery of gravitational waves coming from a blackhole binary coalescence (Abbott *et al.*, 2016).

14.7 Gravitational Waves on Curved Background

So far we have considered only the propagation of gravitational waves represented by perturbations on a flat background. As the real universe consists of matter distribution one needs to consider the situation of the perturbations of Einstein's equations on non flat background. Ehlers and Prasanna (Ehlers *et al.*, 1987; Ehlers and Prasanna, 1996) initiated this first by considering the perturbations of the equations for the case of incoherent dust (p = 0) and later using a WKB formalism for multicomponent fields have discussed the situation both for perfect fluid distribution and for fluids with dissipation like viscosity. A formal treatment of these studies may be seen in the reference (Prasanna, 2017). Summarising the findings it has been found that using the short wave length approximation with the ansatz $U = \exp\left[\frac{i}{\varepsilon}S(x)V\right]$, where the smallness parameter $\varepsilon = \lambda/L$, is the ratio of the wavelength λ to the typical scale length L. The characteristic equation may be obtained from the condition that the determinant of the operator $L_0 = 0$, which for the present case turns out to be, with $T^a = u^a$ the wave vector,

$$(g^{ab}l_a l_b)^2 [(u^a u^b - c_s^2 h^{ab})l_a l_b](u^a l_a)^6 = 0$$

This shows that there are three modes

(i) The gravitational mode, given by the Hamiltonian $H = \frac{1}{2}g^{ab}l_a l_b$, propagating along the null geodesics having the tangent vector $T^a = l^a$.

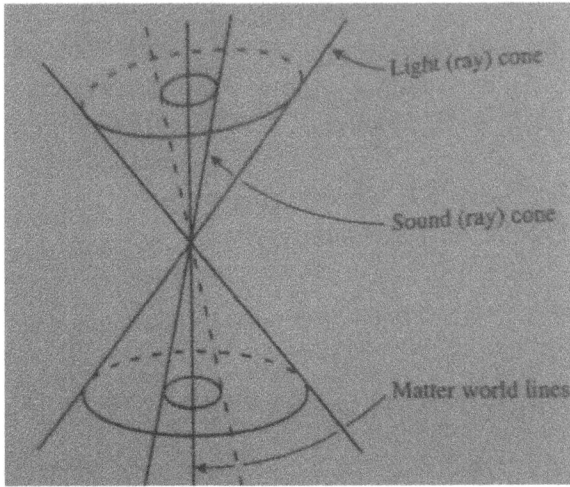

Fig. 14.7. Adopted from (EP 96).

(ii) The sound wave mode given by $H = \left(\frac{1}{2}\right)[c_s^2 h^{ab} - u^a u^b]$, propagating alomg the sound rays with tangent $T^a = \omega(c_s k^a/k + u^a)$.

(iii) The matter mode given by $H = u^a l_a$, moving along the matter rays $T^a = u^a$.

As the matter particles follow geodesics of the unperturbed background spacetime the excitations of the matter mode do not propagate. Figure 14.7 represents graphically the above modes.

It is to be noted that for special cases of matter, situation is as follows:

(i) For incoherent dust ($p = 0$, $\alpha = 0$, $c_s = 0$) there are no sound waves, mode 2 degenerates into the longitudinal part of mode 3.

(ii) For the case of stiff matter $\rho = p$, $c_s = 1$, $\alpha = 0$, sound waves travel with light velocity but are volume changing unlike gravitational waves and M = 2.

(iii) For the case of pure radiation $\rho = 3p$, $c_s^2 = \alpha = 1/3$ and M = 3/2.

Further, it is clear that while the modes for sound waves and gravitational waves are regular, the matter mode is singular.

In this analysis, the transport of amplitudes was also considered through the relation $L_0V_1 + L_1V_0 = 0$. For the lowest order, generally referred to as primary amplitudes, one finds from the first order equations, a set of ordinary differential equations given by

$$(\nabla_l + \theta/2)(a_+^0) = 0 \tag{14.34}$$

and a similar one for a_x^0. Here $\theta = \nabla_a l^a$, $\nabla_l = l^a\nabla_a$. This implies that the change of the complex vector (a_+, a_x) along a ray consists of a rescaling by a positive factor proportional to the square root of the cross sectional area of a small bundle of rays, as in the case of waves propagating in vacua. Further the transport preserves polarisation, helicity and ellipticity alongwith the comserved Issacson stress tensor defined as $\hat{T}^{ab} = (\text{mod} \, a_+^2 + \text{mod} \, a_x^2)l^a l^b/4\pi$.

Considering the transport of the first order primary amplitudes one finds (Prasanna, 2017) for the case of a conformally flat ($C_{hijk} = 0$) spacetime the equation

$$(\nabla_l + \theta/2)a_+^1 + (R/3)a_+^0 = \frac{1}{2}e_+^{ij}[4\nabla_i\nabla^c\delta_j^d - \delta_i^c\delta_j^d\nabla^2 - h^{cd}\nabla_j\nabla_i]v_{0cd} \tag{14.35}$$

which exhibits the possibility of the background curvature as well as the non linear derivatives of the primary amplitudes influencing the transport of the correction to the primary amplitude (Ehlers and Prasanna, 1996).

If one then considers the case of a non perfect fluid with viscosity non zero as given by the stress energy tensor

$$T_{ij} = (\rho + p)u_i u_j + pg_{ij} - 2\eta\sigma_{ij} - \zeta\theta h_{ij}, \tag{14.36}$$

where η, ζ, σ_{ij}, θ, represent the shear, bulk viscosities, shear tensor and expansion scalar respectively. after a lengthy calculation one gets to the final result for the total amplitude $A^2 = 2[\text{mod} \, a_+^2 + \text{mod} \, a_x^2]$ the transport equation to be

$$(D + \nabla^i l_i)A^2 = -2\kappa\eta\omega A^2. \tag{14.37}$$

This final result does show that in a realistic situation where the cosmic medium could be having fluid with disspative forces like viscosity, the propagating gravitational wave could undergo some possible damping, which requires further investigation. (Prasanna, 1999).

After the discovery of gravitational waves by LIGO-VERGO consortium, as one had some numbers about the intensity of the received radiation this study was used to arrive at a possible constraint on the viscosity of intervening dark matter in the path of the arrived gravitational waves. Prasanna *et al.* (Goswami *et al.*, 2017) have studied this effect in the context of the recent observations of gravitational waves. They consider the possibility that the analysis of the GW150914 could allow one to put observational constraint on the shear viscosity if the distance to the source is determined independently. They find that it is in-principle possible to constrain the shear viscosity of the cosmic fluid using GW observations and that the corresponding viscosity values lie in an interesting range which may be relevant to the dissipative dark matter and dark energy models. It turns out that the dissipative dark matter in galaxy clusters such as Abell 3827 (Massey *et al.*, 2015) has the shear viscosity in the range constrained from the GW150914 analysis. Future observations of gravitational waves at LIGO, VIRGO, LISA and other observatories (possibly INDIGO) could potentially probe the viscous properties of cosmological medium (visible and Dark) that pervades our universe.

Chapter 15

Cosmology-the Story
of the Universe

15.1 Introduction

All matter that pervades the space and time is endowed with motion and thus is spread over the vast expanses of space and time. The study of space, time and matter with its motion is referred to as cosmology. The main force or interaction that keeps the motion is gravitation which as described by Newton is universal and sustains mass and energy of the cosmic distribution. It is quite well known that though initially, before the arrival of Copernicus it was thought that the earth was the center of the Universe but with Copernicus the heliocentric theory got established that explained the motion of planets around the Sun by Kepler and with time came the realisation of the motion of Sun in the galaxy, and of galaxy in the local cluster of galaxies and of the cluster in the super cluster so on and on. This vast collection of matter and associated structures, how did they form and how are they holding up has been a question that has occupied the human mind for a long time. However, after the Copernican revolution brought in the heliocentric system of reference to understand the Universe attempts have been initiated to describe the structure of the Universe.

One of the first noticed view of the Universe is it being (looking) the same in all directions for all observers. Geometrically these features are referred to as universe being homogeneous and isotropic. This in fact became the first assumption for constructing any model of the Universe. These two features are actually based on the fact that every observer on earth is in constant motion in space and time as pointed out earlier and thus his/her position in the Universe keeps changing continuously but yet the view of the Universe remains the same. However, as there is no systematic observable motion on a large scale of structures that are major constituents, most astronomers of the nineteenth and early twentieth century assumed the Universe to be static. Before the advent of general relativity, as the dynamics of matter was analysed in the Newtonian framework no model existed to describe the structure and evolution of the Universe. With general relativity describing the matter and the associated gravitational interaction as the curvature of space-time geometry, Einstein (1917) was the first to describe a static universe governed by the space-time metric:

$$ds^2 = dt^2 - \left(1 - \frac{r^2}{R^2}\right)^{-1} dr^2 - r^2(d\theta^2 - \sin^2\theta \, d\varphi^2) \tag{15.1}$$

with the matter distribution satisfying the relations,

$$4\pi(p+\rho) = R^{-2}; \qquad 4\pi(3p+\rho) = \Lambda \tag{15.2}$$

p being the pressure and ρ the density. The constant Λ is called the Cosmological constant. It is worth noting that while developing the theory of relativity, Einstein was influenced by the suggestion of Ernst Mach that the property of inertia that Newton had introduced and is essential for defining the mass of a body is a consequence of the interaction of the body with all other bodies in the universe, which he even termed as Mach's principle. In fact, the equations of general relativity, $G_{ij} = \kappa T_{ij}$, which relate the geometry of space time (universe) with the total content of the energy momentum of matter, in a way reflects this idea as the gravitational potentials g_{ij} solved from these equations, are the basic ingredients of the curvature of space-time. Mach's basic observation was that the velocity and acceleration of a particle would be meaningless if the particle was alone in the universe, as one can only talk of these concepts with respect to other bodies. Further, as the concept of relative velocity led to special relativity, it is the concept of relative acceleration that led to general relativity as it personifies the curvature of space-time. As Einstein was looking for a static model of the universe, and by Newtonian description particles are always attracted to one another, in order to make them static a repulsive force was necessary. He introduced the Λ term in the general equations that provided the repulsive force against gravity, making the material distribution static. Thus, the constant was called the cosmological constant. However, later with Hubble's discovery of the expanding universe scenario Einstein discarded the Λ term saying that 'introducing it was the greatest blunder' he had committed. The story does not end there as with modern cosmology the Λ term plays a very prominent role particularly in the context of 'dark energy', which may be discussed later.

With the introduction of the cosmological constant Einstein's equations of general relativity take the form:

$$R_{ij} - \frac{1}{2} R g_{ij} + \Lambda g_{ij} = \kappa T_{ij} \qquad (15.3)$$

This does not change the conservation law $T^{ij}_{;j} = 0$, as the newly added term also satisfies $(\Lambda g^{ij})_{;j} = 0$, since g_{ij}'s are covariant constants. From equations (15.2) it can be seen that for a universe dominated by incoherent matter (also known as pressure free dust) $\Lambda = 4\pi\rho = R^{-2}$, whereas for a radiation dominated universe with the equation of state $\rho = 3p$, the pressure is $\pi p = (4R)^{-2}$, and $\Lambda = 3/2R^2$.

15.2 Cosmological Models

The mathematical models of the Universe we live in are built on two main postulates: (1) the Weyl postulate and (2) the cosmological principle. The Weyl postulate is mainly to define the cosmic time that can be used to compare observations. According to this principle, the world lines of the particles constituting the Universe (geodesics of the underlying space time structure) are always orthogonal to the time like hypersurface $t =$ constant, thus relating the time on different particles such that at every instant the time is considered as the cosmic time linking all observers. If the Universe seems to show that there was a beginning then all the geodesics emanate from the single point at a distant past (commonly referred to as a singularity) and diverge. On the other hand if the Universe had no beginning or end but always in a steady state then there would not be a singularity.

Milne and McCrea (1934) were the first to give a simple model as applied to Newtonian physics according to which at the time t = constant the Universe is assumed to be homogeneous and isotropic. Satisfying these requirements the only non-zero velocity of particles (constituents) would be the radial component having the angular components constant. Observationally, Edwin Hubble had already in 1928 shown that the light from distant galaxies showed 'red-shift' of spectral lines indicating that the emitting sources were moving away from each other and farther a galaxy is located, it appeared fainter with larger red-shift. Hubble had worked out empirically a relation between the distance and the red-shift which showed that the redshift is directly proportional to the distance. This relation came to be known as Hubble's law as given by $z = DH/c$, with c the velocity of light and H a constant called Hubble's constant. Hubble's first estimate of the constant was 1.5×10^{-17}. The cosmological principle introduces the idea of fixing an origin of coordinates which with the assumption of homogeneity and isotropy could be any point in space as at any instant of time all points bear the same relation of distance and velocity, H and c being constants. As the observational studies increased and more data could be gathered it was found that the numerical value of H changed slightly and the latest finding seem to give a value of 69.8 km/sec/Mpc (Freedman, 2021). Without going into details of these measurements for the present, one may consider the basic aspects of the description of the universe as supported by the space-time picture of matter and the associated geometry as described by general relativity.

Mainly based on the ideas of homogeneity and isotropy of space, it can be easily realised that in three dimensional geometry (t = constant hypersurfaces) spherically symmetric space comes out to be the best idealisation that presents an observer uniformity of view from every point. It is thus the earliest attempts to present a cosmological model (Einstein, 1917), and (deSitter, 1917) both obtained spherically symmetric solutions of Einstein's equations but found different solutions, static in Einstein's case and non-static but stationary in the case of deSitter. As Tolman pointed out later, starting with a general spherically symmetric line element, given by

$$ds^2 = e^\nu \, dt^2 - e^\lambda dr^2 - r^2(d\theta^2 - \sin^2\theta d\varphi^2) \tag{15.4}$$

With λ and ν being functions of the radial coordinate r alone, writing fully the components of the field equation for a perfect fluid distribution including the Λ term and $(T_i^j = (\varrho + p)u_i u^j - p\delta_i^j)$, one finds the equations to be satisfied are

$$8\pi p = e^{-\lambda} (\nu'/r + 1/r^2) - 1/r^2 + \Lambda \tag{15.5}$$
$$8\pi \varrho = e^{-\lambda} (\lambda'/r + 1/r^2) + 1/r^2 - \Lambda \quad \text{and}$$
$$dp/dr = - (p + \varrho)\nu'/2$$

along with the conditions, the pressure p as measured by the local observer shall be the same everywhere, and the proper macroscopic density ϱ also be the same, both satisfying the condition of homogeneity, and that the metric is asymptotically flat and allowing the special theory to hold good at every point in a small neighbourhood satisfying the strong principle of equivalence. If the pressure is to be the same everywhere, one should have either

$$\nu' = 0, \text{ or } \quad p + \varrho = 0 \text{ or both to be zero.}$$

Let us consider the three cases separately

(1) $\nu' = 0$, Including the third condition this gives ν to be a constant as the only solution. This yields in turn a solution for λ to be, $e^{-\lambda} = 1 - (\Lambda - 8\pi p)r^2$. Introducing the constant $R^{-2} = (\Lambda - 8\pi p)$, one can write the line-element to be

$$ds^2 = -dr^2 / \left(1 - r^2 \Big/ R^2\right) - r^2(d\theta^2 - Sin^2\theta d\varphi^2) + dt^2$$

This in fact is one of the forms of Einstein static universe as mentioned earlier.

(2) $(\varrho + p = 0)$, This implies the equation $e^{-\lambda}(\lambda' + \nu')/r = 0$, leading to $\Rightarrow \lambda' = -\nu'$. As both satisfy the same boundary condition, this gives $\lambda = -\nu$. Since the density ϱ has to be a constant, integrating the equation

for λ, one gets the solution, $e^{-\lambda} = 1 + A/r - (\Lambda + 8\pi\varrho)/3)r^2$, A being a constant of integration. Using again the condition that the solution should satisfy the special relativity values for very small r, the constant A is chosen to be zero. Thus, finally one has the complete solution,

$e^{\nu} = e^{-\lambda} = 1 - (\Lambda + 8\pi\varrho)/3)r^2$. Introducing a separate constant R defined as $R^{-2} = (\Lambda + 8\pi\varrho)/3$, the complete line-element for this case turns out to be

$$ds^2 = (1 - r^2/R^2)dt^2 - (1 - r^2/R^2)^{-1} dr^2 - r^2(d\theta^2 - \sin^2\theta d\varphi^2)$$

This solution was obtained by deSitter in 1917.

(3) The third case with both the factors zero yields the special relativity line element as the solution requires both λ and v to be zero.

A few years later in 1922, Friedmann the Russian mathematician obtained a class of solutions, now known as Friedmann cosmological models, which are homogeneous, isotropic and non-static, derived on the basis of the cosmological principle and the Weyl postulate.

One begins with a homogeneous and isotropic $t =$ constant, hyper surface the three dimensional space of constant curvature whose metric is represented by the 3-surface:

$$dl^2 = (dx^2 + dy^2 + dz^2)/(1 + k(x^2 + y^2 + x^2)/4a^2)^2 \tag{15.6}$$

where k/a^2 is the curvature and a the radius of curvature. The values of $k = 0, +1, -1$, indicate zero, positive and negative curvature, identifying a flat space from spaces of constant curvature (positive curvature indicating a spherical surface while the negative curvature stands for a hyperbolic surface). Introducing a change of variable, (rescaling the coordinates $x' = x/a$, $y' = y/a$ and $z' = z/a$), and writing back without the dash, one can write the four dimensional line element

$$ds^2 = c^2 dt^2 - a^2 dl^2, \tag{15.7}$$

$$dl^2 = (dx^2 + dy^2 + dz^2)/\left[\left(1 + \frac{k\left(x^2 + y^2 + z^2\right)}{4}\right)\right]^2 \tag{15.8}$$

where in the coordinates x, y, z. are co-moving or the Lagrangian coordinates. The scale factor a is a function of the free parameter time t. As polar coordinates are easy to handle algebraically, one can express the 3-geometry in the form

$$dl^2 = \left[\left(1 + \frac{kr^2}{4a^2} \right) \right]^2 (dr^2 + r^2 d\theta^2 + r^2 \sin^2 \theta d\varphi^2) \tag{15.9}$$

Rewriting the radial coordinate $\bar{r} = r / (1 + r^2 / 4a^2)$ and again removing the bars the line-element takes the form such that the final form of the 4-metric reduces to

$$ds^2 = dt^2 - a^2 dl^2 \tag{15.10}$$

$$dl^2 = dr^2 / \left(1 - \frac{kr^2}{a^2} \right) + r^2 d\theta^2 + r^2 \sin^2 \theta d\varphi^2 \tag{15.11}$$

generally considered as the standard form for cosmological models.

With these one can now write the field equations for a perfect fluid distribution

$$T_i^j = (\varrho + p)u_i u^j - p\delta_i^j \tag{15.12}$$

for the metric given above to be:

$$T_0^0 = \left(\frac{8\pi G\rho}{3} \right) = (\dot{a}^2/a^2 + kc^2/a^2) \tag{15.13}$$

$$T_1^1 = T_2^2 = T_3^3 = \left(-\frac{8\pi GP}{c^2} \right) = 2\ddot{a}/a + (\dot{a}^2/a^2 + kc^2/a^2)$$

The conservation law (because of contracted Bianchi identities) gives the relation,

$$\dot{\rho} + \left(\frac{3\dot{a}}{a} \right) \left(\rho + \frac{p}{c^2} \right) = 0 \tag{15.14}$$

and the set of equations (15.13, 15.14) are known as Friedmann equations for Cosmology. As only two of this set of equations are independent, one needs an equation of state to solve for all the three variables p, ρ and

$a(t)$. One can right away notice that if both pressure and density are positive, the ensuing equation $\rho + p = -2\ddot{a}/a$ indicates that $\ddot{a}$ has to be negative, showing that $\dot{a}$ has to be positive or negative (not a constant) which consequently implies that the universe must either be expanding or contracting. As the observational data was not there in the early twenties, Friedmann used the condition for a pressure less dust ($p = 0$) filled universe which gives for $a(t)$, the equation:

$$2\ddot{a}/a + (\dot{a}^2/a^2 + kc^2/a^2) \tag{15.15}$$

which can be solved for the three separate cases, $k = 0$ or ± 1.

These set of models were rediscovered by Lemaitre (1927), Robertson (1935) and Walker (1936) using different coordinate systems, which is quite valid due to general covariace of Einstein's equations. The most used form of the solution in modern Cosmology is written as

$$ds^2 = dt^2 - R(t)^2 \left[\frac{dr^2}{1 - kr^2} + r^2(d\theta^2 + \sin^2\theta d\varphi^2) \right] \tag{15.16}$$

and is called Friedmann–Lemaitre–Robertson–Walker (FLRW) line-element (or metric) where again the constant k takes the values 0, +1 or −1. It was shown by Robertson and Walker independently that these are the most general forms of the line element for a spatially homogeneous and isotropic metric, independent of Einstein's equations. Another popular format of these models is obtained by rewriting the radial coordinate $r = \bar{r}\left(1 + \frac{k\bar{r}^2}{4}\right)$ which yields the line-element in the form

$$ds^2 = dt^2 - [R(t)^2 / \left(1 + \frac{k\bar{r}^2}{4}\right)^2][d\bar{r}^2 + \bar{r}^2(d\theta^2 + \sin^2\theta d\varphi^2) \tag{15.17}$$

The function $R(t)$ replaces the earlier used function $a(t)$ the scale factor and the overhead dash on the coordinate r is dropped. Again the constant k takes the values, 0, 1 or −1 according to the spatial curvature being zero, positive or negative. This form indicates the inherent isotropy of the three space as it should be for homogeneity of space and the function Rr indicates the distance function of any point varying with respect to time from the origin of coordinates. As expressed by the Friedmann equations, the pressure and density distribution are expressed through the equations,

$$8\pi G\rho = \frac{3}{R^2}(kc^2 + \dot{R}^2) \text{ and } \frac{8\pi Gp}{c^2} = -2\frac{\ddot{R}}{R} - \frac{\dot{R}^2}{R^2} - \frac{kc^2}{R^2} \qquad (15.18)$$

showing that both pressure and density are independent of spatial coordinates but functions of the time 't' alone. Because of this, one finds that the idealised fluid distribution supporting the geometry is uniform throughout the time slice thus qualifying the model universe under consideration as a uniform model universe and the entire class as *uniform model universes*.

Alongwith these two equations for the parameters of a perfect fluid distribution, and the variation of the scale factor $R(t)$ the Friedmann equations also encompass the conservation law as given earlier. On simplification these equations together yield an equation for R as given by

$$\dot{\rho} + 3(p+\varrho)(\dot{R}/R) = -\frac{3}{8\pi}(\dot{R}/R)[3\dot{R}^2/R^2 + 3k/R^2 - \Lambda] = 0 \qquad (15.19)$$

which on regrouping gives $\dfrac{d}{dt}(\rho R^3) + \dfrac{pdV}{dt} = 0, \quad V = R^3 \qquad (15.20)$

As density is constant over the 3-space t = constant, one can rewrite this as the equation of continuity or conservation of energy $dE + pdV = 0$, which expresses the first law of thermodynamics.

Recollecting earlier attempts of Einstein and deSitter to obtain models of the universe as mentioned earlier (section 1) Einstein modified the field equations with the Λ term in order to get a repulsive force against gravity for obtaining staticity. With the modified equations having the Λ term, for a perfect fluid configuration, one finds for the pressure and density the equations in the Friedmann notations as given by

$$8\pi G\rho = \frac{3}{R^2}(kc^2 + \dot{R}^2) - \Lambda \text{ and } \frac{8\pi Gp}{c^2} = -2\frac{\ddot{R}}{R} - \frac{\dot{R}^2}{R^2} - \frac{kc^2}{R^2} + \Lambda \quad (15.21)$$

It may be seen that these equations also lead to conservation of mass as derived above and thus continue to hold for uniform model universes whether with or without the Λ term. The Einstein solution for the static universe however was later shown by Lemaitre (1927) to be unstable, as obtained from the set of equations $(4/3)\pi R^3 \rho = M$ a constant and $R(t)$ satisfies the differential equation $\dot{R}^2 = 2GM/R + \Lambda R^2/3 - k$. For Einstein static universe, with $k = +1$, Λ takes the specific value $\Lambda_c = (64\pi^2/9k^2M^2)$, with the scale factor R_c taking the value corresponding to Λ_c. As pointed out by Lemaitre (1931), if R is slightly perturbed from this critical value where R_c reaches Λ_c, $\dot{R}$ keeps on increasing or decreasing showing instability. These two simple solutions were referred to by Eddington as one 'universe with mass but without motion' (Einstein) and the other 'universe with motion but without mass' (deSitter). Friedmann considered the case of universe with incoherent (pressure free) dust and obtained a solution of the second order differential equation

$$-2\frac{\ddot{R}}{R} - \frac{\dot{R}^2}{R^2} - \frac{kc^2}{R^2} + \Lambda = 0, \text{ whose first integral is} \qquad (15.22)$$

$$R(\dot{R}^2 + k) - \frac{1}{3}\Lambda R^3 = C \quad \text{a constant.} \qquad (15.23)$$

On using the other equation one can find that $C = \frac{8}{3}\pi R^3 \rho$. Eliminating ρ between (18) and (19) one gets the Friedmann equation

$$\dot{R}^2 = \frac{C}{R} + \frac{1}{3}\Lambda R^2 - k. \qquad (15.24)$$

Replacing the constant C by the mass parameter M, one can rewrite it as

$$\dot{R}^2 = \frac{2GM}{R} + \Lambda R^2/3 - k. \qquad (15.25)$$

which is called the Lemaitre form as mentioned above, and yields the critical scale factor for a static universe $R_c = 1/\sqrt{\Lambda_c}$. It is around this time that Edwin Hubble disclosed his interpretation of the red-shift of spectral lines of galaxies (1930) and the simple thumb rule he found between the luminosity distance and the red-shift $cz = Hd$. Here, z indicates the red shift, c the velocity of light, d the distance and H the proportionality constant, later came to be known as Hubble constant. It implied that the motion in the cosmos is associated to the distances between material particles from each other and with time the material particles (galaxies clusters etc) moved away from each other giving rise to the expansion of the Universe causing the redshift of spectral lines of galaxies. This obviously results in the increasing of distances between galaxies which would mean that the universe is expanding. Further as was known already from the 'Doppler effect', when there is a red-shift of a source of light it is interpreted as the recession of the source (moving away from the observer) with a velocity $v = z\,c$ and thus one can write the Hubble relation $v = H\,d$, v being the velocity of recession and d the distance, showing that the velocity is homogeneous and isotropic as wanted by the cosmological principle. This recession of particles from each other (in this case the galaxies) leads to what is called the 'expansion of the universe' leading to the theory of expanding universe. It is the realisation of this fact that made Einstein reject his own suggestion of the 'cosmological constant' Λ. However we continue to study the features of all possible models with varied values of both constants k and Λ.

Before that we need to introduce two more important parameters the critical density ρ_c and the deceleration parameter q_0. As derived in (Zeldovich and Novikov 1983) the law of expansion $v = H\,d$ leads to the local picture of expansion as given by the rule that the distance between any two galaxies say A and B changes by the Hubble law $dr_{AB}/dt = Hr_{AB}$, which on integration leads to

$$r(t)_{AB} = r(t_0)_{AB} \exp\int_{t_0}^{t} H(t)dt. \qquad (15.26)$$

Considering a sphere of radius $R(t)$ and mass M of particles constituting the system., its density varies as

$$\frac{d\rho}{dt} = \frac{-9M}{4\pi R^4}\frac{dR}{dt}. \qquad (15.27)$$

As dR/dt is $v = HR$ one gets on simplification the rule

$$\frac{d\rho}{dt} = -3\rho H. \qquad (15.28)$$

In fact one could obtain the same relation using equation of continuity $\partial\rho/\partial t = -$div ρv and the law $v = Hr$, as div $r = 3$ and ρ is a function of time only. On any particle on the surface of such a sphere, the acceleration due to the mass of the sphere is given by the relation for the acceleration of a particle $\frac{d^2R}{dt^2} = -\frac{GM}{R^2}$. Using the earlier mentioned relations $\frac{dR}{dr} = HR$, $M = \frac{4}{3}\pi r^3\rho$ and simplifying one can get the equation

$$\frac{dH}{dt} = -H^2 - \frac{4\pi}{3}G\rho$$

(15.29)

The system of equations, (15.27, 15.28) determine completely the changes with time of all the local properties of the universe. Integrating the equation of motion one gets the energy equation: $\dot{R}^2/2 - GM/R =$ constant.

It is useful to express these various inbuilt relations among the measurable quantities in terms of the scale parameter $R(t)$ of the FRW spacetimes as follows (Ray, 1992).

Let us start with assuming the present epoch to be at $t = t_0$, and consider the light emitted from a source at a point P at time t_1, $(t_1 < t_0)$. From the understanding of the light propagation, one knows that the light would have spread over a spherical surface with center P_0 $(t_0, r = r_1)$ and passing through the point, event $O_0(t_0, r = 0)$, whose surface area would be the same as that of a surface centered at O_0 and passing through P_0 as required by the homogeneity of space (Fig. 15.1) (Ray, 1992).

From the associated metric for the space-time (15.15) one can write the line element for this sphere as $ds^2 = -[R(t_0)r_1]^2[d\Omega^2]$, giving the surface area to be $4\pi R(t_0)^2 r_1^2$ following which the observed intensity should be given by

$$I = E/4\pi r_1^2 R(t_0)^2(1+z)^2$$

(15.30)

where E is the energy radiated per unit time by a distant source and I the light intensity received/unit area/unit time. In an expanding universe one has taken into account the Doppler shift factors due to time dialation and energy reduction. As in an Euclidean space the distance of a source is defined in terms of observed or apparent luminosity $(l_d = E/4\pi I)^{1/2}$, and the

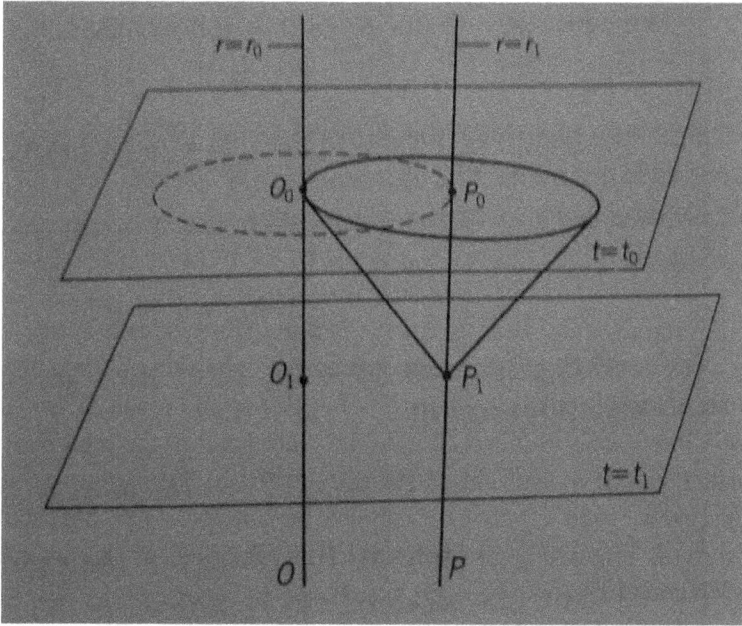

Fig. 15.1. Adopted from Ray (1992).

luminosity distance defined basically as the square root of the ratio of absolute luminosity (E) to apparent luminosity. Using the above definitions one can express the 'luminosity distance' d_L in terms of the scale factor $R(t_0)$, $d_L = r_1 R(t_0)$. Since the concepts of distance, velocity and the red shift z are all related through Hubble relation, one can see that the Hubble parameter $H(t)$ can be expressed as $H(t) = \dot{R}(t)/R(t)$, giving the Hubble's law in relativistic cosmology to be $z \cong H(t_0)d_L$. The deceleration parameter q is defined as $q = -R\ddot{R}/\dot{R}^2$.

As derived earlier the pressure and density in the Friedmann model are given by the equation (15.18) and from that one can see the relations ($c = 1$, units)

$$-4\pi G(\rho + 3p)R = 3\ddot{R} \quad \text{and} \quad -4\pi G(\rho - p)R^2 = R\ddot{R} + 2(\dot{R}^2 + k) \quad (15.31)$$

Eliminating $\ddot{R}$ from these two subequations, one gets the equation for $\dot{R}$, $\dot{R}^2 + k = 8\pi(G\rho/3)R^2$, along with the equation of energy conservation

$$\frac{d}{dt}\left[R^3(p+\rho)\right] = R^3\dot{p} \tag{15.32}$$

Another form of conservation law that comes from field equations yield equation (15.14) where the function '*a*' is to be replaced by $R(t)$ thus giving

$$\dot{\rho} + 3(\rho + p)\dot{R}/R = 0 \tag{15.33}$$

From the existing data one knows that the Universe had a very early phase when basically radiation dominated the physical content when the equation of state would have been $\rho = 3p$, that would result in the above equation integrating to $\rho = C_1/R^4$. On the other hand in the present epoch the matter has dominated the physical content and the matter density is extremely low, (supported by observations) and the pressure is almost taken to be zero (as assumed by Einstein and later by Friedmann) and this results in the equation of state to be $p = 0$, which integrates the conservation equation to $\rho = C_2/R^3$. This argument can easily show that in the very distant past universe was radiation dominated and with time and as the scale factor R got bigger the matter started dominating allowing the universe to cool faster than the earlier epoch. In fact this feature lead Lemaitre (1927) to predict that the early Universe must have been in an extremely hot and dense phase with the density almost infinite before it exploded in a huge way letting out some matter but mostly radiation in a '**big bang**' (a comment mockingly put forth by Fred Hoyle). As radiation cools faster than matter, universe started cooling faster in a phase, later described by Gamow *et al.* (1948).

It may be seen from (15.30) that at the present *epoch* $t = t_0$, the density and pressure are given by

$$\rho_0 = (3/8\pi G)(H_0^2 + k/R_0^2) \tag{15.34}$$

and

$$p_0 = -\left(\frac{1}{8\pi G}\right)\left[H_0^2\left(1 - 2q_0\right) + k/R_0^2\right] \tag{15.35}$$

With $R_0 = R(t_0)$, H_0 and q_0 being defined earlier. It may be seen from (15.34) that the curvature constant k should be positive or negative depending upon whether the density ρ_0 is greater or lesser than the value $\rho_c = 3H_0^2/8\pi G$, called the 'critical density'. As pressure p_0 is considered to be zero at t_0 one gets the relation $k/R_0^2 = H_0^2 (2q_0 - 1)$ showing that together the relations give the ratio of the present density to the critical density to be,

$$\rho_0/\rho_c = 2q_0 \tag{15.36}$$

15.3 Classification of FLRW Models

As was seen in the beginning, these models are in principle characterised by the parameters Λ and k apart from the physical parameters like z the red shift, the density ρ and pressure p. However, in the above discussion, we did not consider the role of Λ which we will now take up in the discussion.

Case 1, $\Lambda = 0$. This refers to the flat space or the Euclidean 3-space. The Friedmann equation (15.23) reduces to

$$R(\dot{R}^2 + k) = C \tag{15.37}$$

and would require to be studied for different values of k

(1) $k = 0$
Equation reduces to $\dot{R}^2 = C/R$, which on integration with the boundary condition that $R = 0$, when $t = 0$, gives the solution $R = (9/4 \, Ct^2)^{1/3}$. The model is named after Einstein and deSitter and called Einstein–deSitter model, also referred to as the critical model as it lies between the open models ($k = -1$) and closed models ($k = +1$) This model implies that $q_0 = 1/2$ as k is zero and thus from (15.36) one finds $\rho_0 = \rho_c$, and as depicted in the figure $R(t)$ increases as given by

$$R(t)/R_0 = (3H_0t/2)^{2/3}. \tag{15.38}$$

Further one can from earlier results find the age of the universe $\approx(H_0^{-1})$ in this epoch to be $\approx 9 \times 10^9$ yrs (Weinberg 1972).

(2) $k = +1$

The equation is $\dot{R}^2 = C/R - 1$, whose solution may be written as

$$t = C\left[\sin^{-1}(R/C)^{\frac{1}{2}} - (R/C)^{\frac{1}{2}}(1 - R/C)^{\frac{1}{2}}\right] \tag{15.39}$$

the equation to a cycloid thus showing an oscillatory character for the universe with an elliptically closed 3-surface. The scale factor behaves such that the expansion starts from a point $r = 0$, generally referred to as a 'singularity' reaches a maximum and starts contracting to a singular point. The proper density in this case $\rho_0 > \rho_c$

(3) $k = -1$

The equation is $\dot{R}^2 = C/R + 1$, whose solution is

$$t = C\left[(R/C)^{\frac{1}{2}}(1 + R/C)^{\frac{1}{2}} - \sinh^{-1}(R/C)^{\frac{1}{2}}\right] \tag{15.40}$$

showing that the scale factor increases without limit as the function is given by

$R(t)/R_0 = [q_0/(1 - 2q_0)]\,(\cosh\theta - 1)$, tends to $(1 - 2q_0)^{1/2}\,H_0 t$

as t tends to *infinity*. This is called the ever expanding Milne(1931) universe with the 3-hyper surface being open and hyperbolic.

Figure 15.2 gives a brief sketch of the scale factor as a function of time for these three classes of models.

Case 2, $\Lambda > 0$. There is a particular value of $\Lambda = \Lambda_E$, which is of interest for all the cases of the constant $k = 0, \pm 1$ For $k = +1$, the plots of R against t are depicted in Fig. (15.3).

(i) As can be seen in this case, $0 < \Lambda < \Lambda_E$, yields a bouncing model with the behaviour depending upon the distant past as it shows the contracting phase reaching a minimum and then going to expansion phase. However, if there was a beginning for the universe starting from a singularity, then the upper curve is ruled out with the universe reaching a maximum radius and then going for a crunch.

(ii) If $\Lambda = \Lambda_E$, we have the Einstein model, with the gravitational attraction being opposed by the Cosmic repulsion due to Λ as indicated by a flat line.

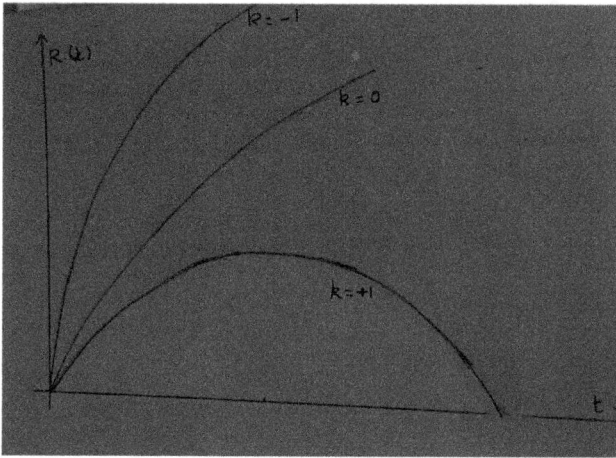

Fig. 15.2.

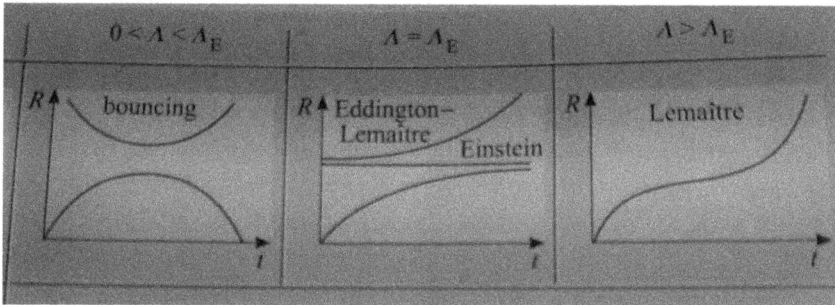

Fig. 15.3. Adopted from RDI 92.

(iii) If $\Lambda > \Lambda_E$, one gets the Lemaitre model, which is indefinitely expanding, called the Eddington–Lemaitre model that seems to start either from a point singularity or static Einstein universe and because of some perturbation starts expanding, asymptotically approaching $R \approx \exp[(\Lambda/3)^{1/2}t]$.

Case 3. One then looks at the possible scenario for the case $k = 0$, and $\Lambda > 0$

This yields an ever accelerating universe with acceleration increasing with time and the radius asymptotically reaching the same value as in subclass (iii) above with $R \approx \exp[(\Lambda/3)^{1/2}t]$. It starts with a singularity and

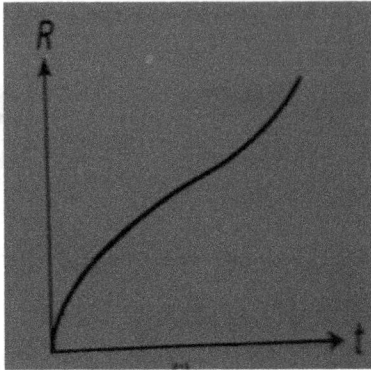

Fig. 15.4.   Adopted from Ray (1992).

expands continuously with a slight slowing down at some phase but then increasing the acceleration. This is the favoured model today as we will get to the details later.

What happens if the cosmological constant is negative? It is a little surprising that irrespective of the curvature constant (0, +1 or −1), one finds the model to be bouncing after the initial expansion but then after reaching a maximum turns back to shrinking which again bounces to the stage of expansion and this cycle seems to repeat itself yielding an *oscillating universe*.

Summarising this discussion, it may be seen that in the absence of the cosmological constant ($\Lambda = 0$), $k = +1$ leads to what is called the 'closed model' with the property that at $t = 0$, R is zero and with time the universe expands but then with advancing time, the model shows a contracting phase that ends in a crunch. The model seems to advance the idea that the Universe starts with a sudden expansion from a singularity and seems to end in a singularity and thus being referred to as a closed model. On the other hand the case when the curvature constant $k = -1$, leads to ever expanding universe and appropriately called the 'open model', as the scale factor R varies proportional to t. The other scenario with $k = 0$, leads to what is called the 'critical model' with the expansion factor R being proportional to $t^{2/3}$ for all time. As this model received support from both Einstein and deSitter, this is also referred to as Einstein–deSitter Universe. With $\Lambda = \Lambda_E$, one finds depending upon the constant of integration three distinct behaviours,

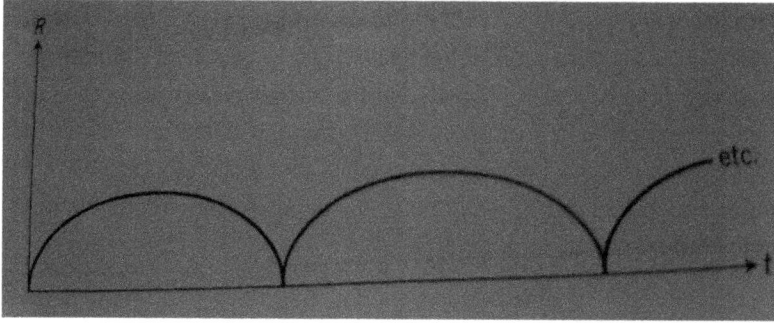

Fig. 15.5. Adopted from Ray (1992).

(i) Einstein static Universe with the gravitational attraction being coun-
terbalanced by the cosmic repulsion,
(ii) Instant initial expansion (popularly called the BigBang) finally leading to Einstein static scenario, and
(iii) Einstein–Lemaitre model where running back in time one approaches
the static Universe and in forward direction it is ever-expanding.

During the period 1925 to 1935 when Lemaitre proposed the idea of
initial explosion leading on to ever expansion ($k = +1$, $\Lambda > \Lambda_E$), it attracted
a lot of attention as it evoked interest in the origin of chemical elements
actually the nuclei, due to the hot dense phase of the model, and thus
became practically the most accepted model later.

The model starts with $R = 0$, at $t = 0$, and passes through a pseudo-
static phase and then expands continuously. The pseudo static phase could
have been the era for formation of stars and galaxies with the later expansion bringing the presently observed Universe (Jones *et al.*, 2005).

The three distinct periods of this model (Lemaitre 1935) may be dis-
tinguished as (Ray, 1992)

(i) a period of expansion from a point source during which the chemical
elements were formed,
(ii) period of reduced expansion favourable for the formation of structures and finally,
(iii) the period of renewed expansion during which the recession of the
nebulae is accelerating with the unlikely possibility of new condensa-
tions because of the diminished density.

It is thus by the end of 1930s that general relativity got a prominent position for describing cosmology as a science with observational parameters being analysed theoretically and a scenario well developed. The Lemaitre model took a prominent position with it being recognised as the big bang model which later got established along with the early Universe scenario as worked out by Gamow *et al.* in the forties including the elemental synthesis, getting popularly known as the 'Big bang theory'.

15.4 A Slight Historical Diversion

Recalling the early mention of the theoretical premise used for cosmology-the cosmological principle one knows that this principle ascertained that at any given time the spatial slice of the universe appeared the same for all observers-a direct consequence of homogeneity and isotropy. In 1948, Bondi and Gold proposed to study the extension of the cosmological principle to 'perfect cosmological principle which said that the universe appears the same for all observers at all times, independent of the era and thus it is in a *steady state*. In 1949, Fred Hoyle proposed to modify the equations of general relativity, particularly the stress energy tensor to include creation of matter depleted by the expansion of space thus keeping the density constant to achieve the same steady state formulation. These modifications lead them to propose a theory later called the steady state theory and it provoided another approach for cosmology as against the big bang theory where the density keeps decreasing with age. However, both the approaches use the Weyl postulate and general relativity for space-time description as required by any cosmological model. Finally it is the observational evidences properly explained that will judge the most accepted theory of the Universe. As we have already touched upon the aspects then known in the 30s which later found to be supporting the Big bang theory, it is of scientific interest to sketch the considerations of Bondi, Gold and Hoyle in the late 40s and later by Hoyle and Narlikar in the mid sixties for proposing the steady state model.

It is very fortunate that Radio astronomy developed in the forties and within a decade became a very useful tool. After Hubble's discovery of the galactic recession through observing their red shifts the initial attempts for interpreting cosmic distribution was to check the correlation between the red shifts vs number counts of galaxies. As the number of galaxies in the far away distance was enormous the project was given up. However, in the mid 50s with the realisation that the number density of radio sources is considerably lower than that of galaxies it was thought that plotting the number of radio sources against their flux densities (logN vs. logS) might be more informative. Sir Martin Ryle was the leader in this venture. The main contention in this approach was to find the slope of the (logN vs. logS) curve which according to Euclidean space conjecture should be flattening for an expanding Universe scenario (zero curvature) reducing the slope of the curve as one gets from higher to lower flux density. On the contrary the data of Ryle showed just the opposite with slope increasing to 3,0 from 1.5 the Euclidean value. These initial studies however seemed to show that the steady state approach had a better support from observations as compared to the expanding universe model from a point. Further improved survey of radio sources by Ryle's group (4C catalogue) saw the slope reduction to 1.8 due to addition of many more sources and thus the contention between the expanding vs steady state models continued. Can observational features provide any clue that tries to distinguish between these models? One of the main features that could bring a difference is the average density which the steady state requires to be constant whereas the expanding big bang model allows it to decrease with age. It was already seen that the energy density of rest mass varies as $1/R^3$, whereas that of radiation as $1/R^4$ a fact that encourages one to conclude that in the early stages the expansion of the universe must have been governed by the non relativistic matter content. This lead one to assume that empirical relations between physical parameters like red shifts, luminosities, numbers, angular diameters, etc can reveal only the features of the matter dominated era (Weinberg, 1972). An important input that requires attention concerns the light element formation in the early universe. This study is what is known as 'Hot big bang cosmology', which is considered briefly in the following.

15.5. Early Universe and the Hot Big Bang Model

While discussing cosmological models, it was pointed out how different combinations of the Cosmological constant Λ and the curvature constant k lead to different situations and as was pointed out the most accepted model seems to be the one proposed by Lemaitre in 1929 with $k = +1$, and

$\Lambda > \Lambda_E$, which is also called as Eddington–Lemaitre model. It is mentioned that Lemaitre before adopting the model had also considered the fact that the entropy of the Universe increased after the explosion indicating that before the bang the entropy should have been the minimum indicating fully organised matter collection which was scattered after the explosion. Gamow in 1948 while trying to explain the elemental formation in the universe thus used the super condensed state saying that both the density and temperature at that stage must have been maximum thus providing conditions for elemental formation, which was the basic idea of the so called alphabetical (α, β, γ) paper authored by Alpher, *et al.* (1948). Going back to the earliest assumption for any model universe, *viz* the uniformity (homogeneous and isotropic) one could ask what evidence was there for this assumption. Though at that time (pre 1965) there was no direct evidence, as the observations improved, the observations of visible galaxies suggested that the isotropy of them suggested only upto 30% and this further decreased to ~25% with the addition of radio galaxies. When pervaded with observation of X-rays which was cosmic the isotropicity reduced to a low 5%. But the best observation that phenomenally increased the uniformity came accidentally in 1965. It was like in the case of discovery of cosmic radio signals, the Bell labs had constructed a special antenna intended to use for Echo communication satellites, which was used by Penzias and Wilson, who noted in 1964–1965, that the antenna picked up micro wave signals coming from all parts of the sky, that was just about 3^0 K above absolute zero (Penzias and Wilson, 1965). As they had no idea as to where the radiation was coming from discussion with colleagues spread to theoreticians and Peebles (1965) advanced the idea that this must be the remnant of the 'vanished brilliance' or the fireball of Lemaitre that Universe started with, which Gamow *et al.* (1948) had predicted to appear around 7.3 K. after the cooling due to expansion. Further investigations and observations concluded the radiation as the background radiation pervading the entire universe and consequently showing the assumption of homogeneity and isotropy of the universe to be right and it was named the **'cosmic microwave background radiation'** popularly known as 'CMBR'. Further details of the discovery, properties and significance will be taken up in the following.

15.6 Cosmic Microwave Bacground Radiation

Unlike in the case of radio astronomy, a special antenna built by the Bell telephone company for communication via Echo satellites were being used for a survey by Penzias and Wilson, came across a static emission coming from all directions while trying to measure the intensity of radio waves emitted from our galaxy at high galactic latitudes, out of the plane of the Milky way. It was a difficult proposition as the signals were mostly like noise one hears on a radio set during a thunder storm which is difficult to distinguish from the electrical noise that is produced by the random motion of electrons within the structures of the radio antenna and the amplifier circuit coming from the atmosphere. Generally this type of noise is eliminated by switching the antenna direction back and forth a suspected source. As Penzias and Wilson were attempting to measure the radio noise coming from the galaxy it was very important for them to identify the electrical noise that might be produced within the antenna system. As earlier attempts had revealed a little more noise that could be accounted for they devised a method called 'cold load' by comparing the power coming from the antenna with the power produced by an artificial source cooled with liquid helium about four degrees above the absolute zero. This allowed the cancelling of the antenna power and one expected a very little electrical noise within the antenna structure. In order to check the assumption Penzias and Wilson started their measurements (observation) at a relatively shorter wavelength of 7.35 cm, where the radio noise from the galaxy should have been negligible. Any other possible source from the atmosphere would have had directional dependence as also on the thickness of the atmosphere like being less towards the zenith and more towards the horizon. However, by the spring of 1964, they realised that they were receiving sizeable amount of 'microwave noise at the frequency they were measuring in and was also homogeneous (static noise) without dependence upon the time of the day or season and further was isotropic. This convinced the observers that this radiation as shown by its uniformity and isotropy could not be coming from the Milky way galaxy but from much beyond. After eliminating all possible sources one could think of they realised that it should be coming from outside the solar system and not from any local source but were not able to settle on the origin of this microwave source. However, they calculated the equivalent temperature for this radiation and came to the conclusion that it was about 2.5–4.5 K above the absolute zero. As the news of this observation spread among the radio physicists through private conversation and phone calls, Burke of MIT realised that Jim Peebles of Princeton had been perhaps talking of this through his theoretical studies and the earlier idea suggested by Gamow *et al*, that there could be a left over *remnant of radiation from the initial expansion (big bang) indicating the beginning of the Universe at a very high temperature and emitting γ rays which could have cooled down with expansion of the Universe to a low temperature and the radiation being red shifted to the Far infrared wavelengths.*

This feature should have to be supported by looking at the thermal history of the early universe. The most important property of electromagnetic radiation is its character of having duality. What does this mean? Light was described by Newton as a set of particles (corpuscles) whereas Huygens described it as waves. After electromagnetism and its mathematical description was discovered by Maxwell, it appeared that Light behaves both as waves (phenomenon of interference) and corpuscles (Newton had expressed the view that the phenomena of reflection and refraction are due to its corpuscular nature) Though there were several more features of light that are amenable to describe the wave nature of light, one had to wait till 1905 before Einstein described the photo electric effect as coming from particle nature of light which appeared as packets of energy as proposed by Planck (1900) and termed it as a photon (quantum of energy). As Maxwell had described that light is a form of energy all these ideas got fit into a consistent frame work which also formed a sound basis for the development of quantum theory by Bohr and his colleagues clearly expressing the dual nature of light. As a further support de Broglie's discovery of the wave nature of matter clearly explained the dual nature of both matter and energy explained in a sense Einstein's discovery of the equivalence of mass and energy ($E = mc^2$). With this one can think of radiation as being made of photons with the energy of an individual photon depending upon the frequency of radiation which was the relation enunciated by Planck, $E = h\nu$, h being the Planck's constant. A beam of light therefore can have more energy than another by having more number of photons or lesser number having higher energy. Thus one has an electromagnetic spectrum consisting of the range from radio waves (lower energy) to γ rays (higher energy) differing both in wavelengths and frequencies according to $c = v\lambda$, c being the velocity of light.

The accidental discovery of Penzias and Wilson (1965) of the microwave radiation and its interpretation by Peebles and Dicke (1965) can thus be considered as the beginnings of the study of early universe leading to modern cosmology (1965–66) and the introduction of the idea of the Hot Big bang Universe.

As pointed out by Kolb and Turner (1989), the FRW cosmological model's direct evidence supporting as a probe for conditions in the early universe the validity extends back to the epoch of primordial nucleosynthesis which could have occurred about one hundredth of a sec after the big bang. They have mentioned that sensible speculations could be made to the earliest times, known as Planck scale (10^{-43} secs) after the big bang, made possible by the current standard model of particle physics (Grand Unified Theory). Expansion of the Universe as suggested by Hubble's discovery of the red shifts of galaxies, further supported by those of Quasars shows the universality of the expansion and this interpretation lead to an estimate of the probe to the history from almost about a billion years from the bang (Kolb and Turner 1989). An important feature in this context is the homogeneity and isotropy of the universe which has been made tractable because of the fact that the metric of the model depends upon only one dynamical variable-$R(t)$ the cosmic scale factor. The observed uniformity of the temperature of the CMBR, apart from a small dipole anisotropy caused by our motion with respect to the cosmic frame, is the best evidence for isotropy of the universe. Inhomogeneities in the density of the universe on the last scattering surface (recombination era) would also lead to temperature anisotropies and thus the observed uniformity of CMBR stands for the accuracy of the degree of precision ($\sim 10^{-4}$) to which the homogeneity and isotropy of the universe prevails. As the Hubble parameter is of dimension $1/T$ the present universe is characterised by the time scale

$$H_0^{-1} = \frac{1}{h}\left(\frac{1}{100}\right) \text{sec Mpc/km} = h^{-1} 3 * 10^{17} \text{ sec} \approx 1.4 * 10^{10} \text{yrs}$$

with distance scale given by 4.3×10^3 Mpc. Putting it roughly all distances in the universe will become twice larger in about 10 billion years with galaxies at distance of order 3 Gpc from us moving away with almost equal to the light velocity. As suggested by Gorbunov and Rubakov (2011) irrespective of the cosmological data there exist observational lower bound on t_0 the age to be greater than about 13 billion years. This number is gotten from the observation of the distribution of luminosities of while dwarf stars as they slowly cool down and get dimmer. Another estimate comes from the study of radioactive elemental abundances in the core of the earth and in metal poor stars. Homogeneity and isotropy does not mean that the 3-space is Euclidean as the universe has small spatial curvature as seen from the 3-sphere with positive curvature and the 3-hyperboloid with negative curvature and thus only to a good approximation Universe may be treated as having the 3-space flatness. This as observed from the temperature anisotropy of the CMBR gives bounds that the radius of spatial curvature is much greater than the size of the observable part (H_0^{-1}). As seen from observations the present universe is filled with non interacting cosmic microwave background photons with its number density being about 400 /cc having thermal energy distribution of Planckian spectrum as depicted in Fig. 15.6. The temperature of the photons coming from different directions on the celestial sphere is the same for better than 1/10,000 (modulo dipole component). The angular anisotropy as depicted in Fig. 15.7 is of the order $\delta T/T_0 \sim 10^{-4} - 10^{-5}$ (Gorbunov and Rubakov, 2011). When the Universe was very young and hot, radiation could not have travelled very far without being absorbed and reemitted by some material particles and this continued interaction with energy exchange would have maintained thermal equilibrium resulting in a thermal spectrum of the black body type. After the discovery of CMBR, the COBE satellite's observations confirmed the temperature to be about 2.726 K which was the result of universal expansion. It is important to note that in the very hot and dense early universe thermonuclear reactions produced elements not heavier than hydrogen-deuterium, helium and lithium and the abundance of the computed mix of these elements seem to agree with observations.

These early results came from the studies of Gamow and his associates in 1947. Their work in a sense combined the research in nuclear physics and cosmology. The growth of structure in the early universe was prevented by radiation pressure which changed as the universe cooled to

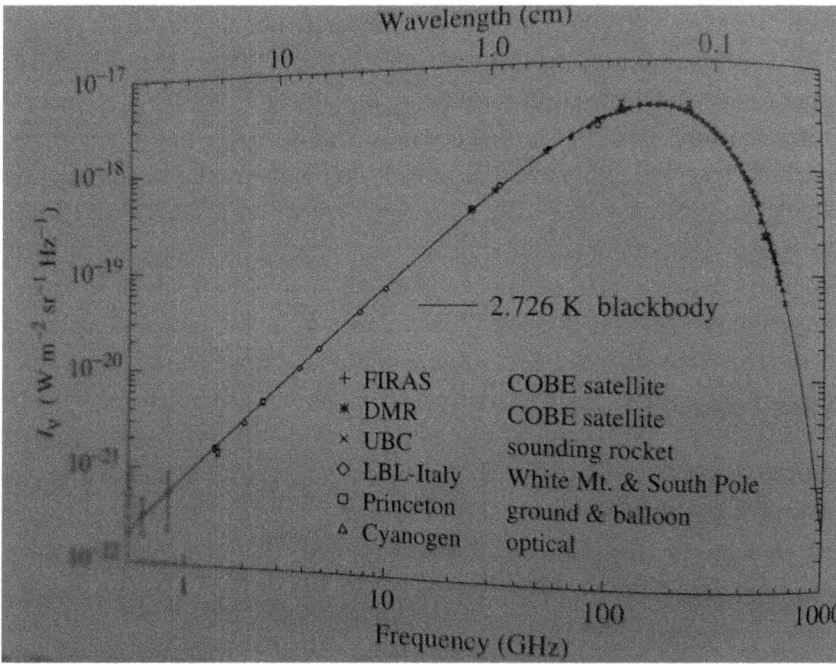

Fig. 15.6. Adopted from Grobunov and Rubakov (2011).

Fig. 15.7 Adopted from Grobunov and Rubakov (2011).

about 3,000 K, when the universe had reached to about 0.1% of its present size. This condition of the universe was cool enough for the recombination of ions and electrons to form neutral atoms of H and He. The neutral atoms so formed lead to gas clouds which could collapse into star clusters. Such clusters lead to a small scale anisotropy in the background radiation which has been observed. The anisotropy shown in CMBR through the warmer regions of the radiation in directions towards clusters is known as Sunyaev–Zeldovic effect (Sunyaev and Zeldoich, 1980) which in fact is measurable. The CMBR as seen today comes from the recombination era of the Universe when the temperature was about 0.3 ev and after recombination the neutral gas was transparent to photons. Further details concerning the hot big bang universe and the associated Inflationary models with its accompanying physics may be found in several standard books and for a non mathematical account of these ideas and their significance one may refer to the following: (Weinberg, 2007; Grobunov and Rubakov 2011; Kolb and Turner, 1989).

The big bang nucleosynthesis that could have occurred at $T \approx 80$ kev. woild have produced only Deuterium, Helium 3 and Lithium 7 and not any higher nuclei. All the heavier elements could have been produced only in stars as the required densities would only exist within stellar configurations. One of the important questions in early Cosmology deals with the baryon asymmetry (baryons are far more abundant than anti baryons) as well as excess of photons per baryon, both of which seem to be the result of the asymmetry in the population of quarks and anti quarks the exact reason for which is still uncertain. The so called Sakharov conditions concerning the baryon asymmetry are (i) non conservation of baryon number, (ii) C and CP violation and (iii) thermal inequilibrium. Though Sakharov was the first to outline a mechanism (Sakharov, 1967) for producing this asymmetry, a complete understanding of the reasons for the phenomena seems to be a still open question. (Grobunov and Rubakov, 2017; Ryden, 2017).

15.7 The Missing Mass and the Dark Matter

The matter that came from big bang nucleosynthesis as known is mostly baryonic in nature. In the last 30 odd years it has been almost clearly interpreted from observations and deductions that the Universe as a whole

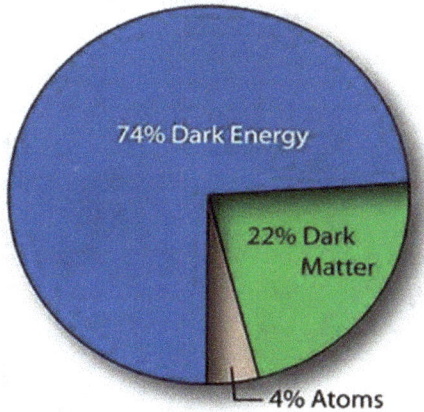

Fig. 15.8. Wikipedia.

may be pictured as in the following Piechart of matter and energy both observed and inferred (Fig.15.8).

It might surprise one to note that almost 74% of energy and matter comes under unseen but deduced from their indirect influence on the observed objects. Even among the 26% of the rest, 22% seems to be considered as 'dark matter' while only about 4% seems to be considered as baryonic matter which came from the nucleosynthesis both primordial and stellar generated nucleons.

The earliest indications of the presence of unseen matter came from the findings of F Zwicky in 1930s (Ostriker, 1999) who, inferring from the then known masses of luminous objects, came to the conclusion that the mass of all stars in the coma cluster of galaxies provided only about a percent of the mass needed to keep the stars from escaping the cluster. This finding was later confirmed in 1970 by Vera Rubin and W. Kent Ford (Rubin, 1988) through observations from other clusters too and further they showed that the mass of stars visible within a typical galaxy is only about ten percent of that required to keep the stars revolving around within the galaxy without escaping. From the studies of the rotational curves of galaxies it was surmised that the speed with which the stars orbit the center of their galaxy -the orbital velocity is either constant or increases slightly with distance from the center and do not follow the Newtonian feature $GM(r)/r = v^2$. Further it was found that the observation

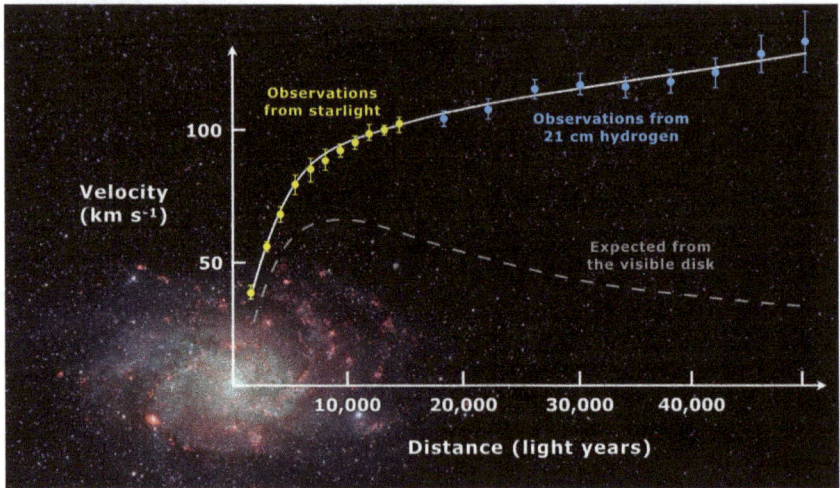

Fig. 15.9. Typical galactic velocity curves. The upper-solid white-line curve is the observational data from galactic starlight (yellow data points) and radio-astronomy spectral analysis (blue data points) of the observed speed of galactic portions at various radial distances out from the galactic center of a typical galaxy such as Messier 33. The lower dashed-grey-line (speed) curves exhibit the expected speed, at the same radial distances. Astrodynamics (Baker, 1967).

of the 21 cms emission from neutral Hydrogen gas clouds in the outer regions of galaxies showed increasing mass $M(r)$ as depicted in the following figure depicting the rotation curves for the galaxy $M33$.

In any given cluster, the random motion of its constituent galaxies tend to disperse the cluster which is balanced by the gravity which would have caused the galaxies to fall towards the center without relative motion. Zwicky who measured the Doppler effect of the galaxies in the cluster found that they were moving so fast that the cluster should be flying apart, and in order to overcome this he concluded that there must be a large amount of unseen matter that keeps the galaxies as they are. This he called the 'Missing mass' which from later studies and confirmations has been called the 'Dark Matter'.

As one can surmise with the masses of galaxies and clusters one may find the average density of the universe estimating the number of galaxies and clusters in a given volume. From the existing data of the distribution of galaxies and clusters and their estimated masses the average density of the visible matter in the universe seems to be about 2.10^{-30} gms/cc. The correct estimate of the average matter density has cosmological implications as it determines the curvature of the universe which is crucial to determine whether the universe is closed and finite or open and infinite. There are several ways of measuring the mass of the Universe. One of the dynamical ways as suggested by Kolb and Turner (1990) is by determining the number density of galaxies times the average mass density of a galaxy and calculating the density parameter $\Omega_0 = <\rho>/\rho_c$. Here, the galactic mass is obtained through the welknown Newtonian relation $GM(r) = v^2r$, v being the orbital velocity of any typical star in the galaxy and r 'the radius of the sphere' surrounding the mass contained within. Using the rotational curves of galaxies and considering that all the light emitted by the galaxy comes from within a sphere of radius r, one can define the Ω_L which turns out to be of the order ≤ 0.01 which is less than about 1% of the critical density. This was disturbing as observation of the 21 cms emission from neutral hydrogen showed a continuous increase in $M(r)$ almost proportional to r. This unobserved mass was the first evidence of unseen or dark matter which seems to create the halo around the galaxy. Continued observations of this feature ensured that all spiral galaxies have diffuse halo which could contribute more than ten times the mass of the visible matter giving the ratio $\Omega_{Halo} \geq 0.1 \cong 10\Omega_L$. This indicates that there could be a substantial amount of dark(unseen) matter in the halo which could be traced to something one is familiar. The ideal way to recognise this is to use the fact that whatever be the form of matter it cannot escape being gravitationally active and thus as a lump of such dark matter passing in front (line of sight) of a known shining object should cause lensing effect by deflecting the light of the background object and such a lensing as already mentioned is termed as microlensing. Such an unseen object is called Massive Compact Halo object (MACHO). If light from a star in the neighbouring galaxy (LMC) is deflected by a MACHO of the Milkyway along the exact line of sight then the Einstein radius of lensing object would be

$$\theta_E = \left[\left(\frac{4GM}{dc^2}\right)(1-x)/x\right]^{1/2}$$

With M being the mass of the MACHO, d is the distance to the lensed star and xd the distance to the MACHO from the observer with ($0 < x < 1$). As was pointed out in an earlier section, if the MACHO is not in the perfect line of the star and the observer, then instead of a ring one will find two or more broken arcs. However, due to the non observation of such events due to the very smallness of the ring as also perhaps the scarcity of objects like brown dwarfs or Jupiter sized planetary objects the general feeling is that the dark matter in the halo may not be baryonic but could be in the form of spread out smooth distribution of non-baryonic dark matter. Among the candidates for non baryonic dark matter, massive neutrinos were the first to be thought of which could have developed a cosmic neutrino background when the universe became opaque to neutrinos during the thermal history when universe was hot and dense enough for the Universe to be opaque to neutrinos. However, given the insufficient mass density of neutrinos suggestions have been made for several alternatives for the role of dark matter. Apart from MACHOS which are baryonic, prominent amongst the non baryonic dark matter are WIMPs which are weakly interacting massive particles. Though several experiments have been performed for detecting wimps so far no convincing detections have been made. Though there is no formal definition of a wimp they interact only gravitationally and do not come under the category of standard model of elementary particles. Wimps are supposed to be relics from the early universe when all particles were supposed to be in thermal equilibrium. At sufficiently higher temperature as was in the early universe the dark matter particle anti-particle pairs would have produced as also annihilated forming lighter particles which process would have ceased with the expansion and cooling of the universe. As the process of dark matter production and pair annihilation would result in gamma rays the exploration for them are concentrated among the gamma ray telescopes. It is also surmised that as the wimps from halo pass through the Sun, they interact with solar protons and heavier nuclei and the neutrinos ejected in such interactions would be of higher energies and possibly may get detected in earth based detectors (Wiki wimp).

Structure formation seems to suggest that from a smooth early universe scenario the density irregularities then developed into structures as seen today. The inflationary hypothesis triggered by the phase transition is understood to be the reason for rapid expansion due to the enormous energy release, where the size of the universe expanded by almost a factor of 10^{50} in a span of less than 10^{-5} secs. As mentioned in the beginning the important ingredient to understand the fate of the universe is the average density vis a vis the critical density which governs the fate of the universe either to expand for ever or be gravitationally bound and be finite. As the amount of baryonic matter is a very small fraction of the total mass and the inclusion of the dark matter also does not improve the ratio to more than one fourth of the total mass and falls short of the critical density as referred to by the inflationary scenario one is bound to think that the Universe may expand for ever and the observational evidence as obtained in the late nineties indicate that the rate of expansion is indeed accelerating. The accelerated expansion of the universe was discovered in 1998 by two independent projects, the Supernova Cosmology Project and the High-Z Supernova Search Team, which used distant type Ia supernovae to measure the acceleration. This information came through the observation of supernovae in distant galaxies (1997) and their redshifts which indicated that the recession of the hosting galaxies of these supernovae is at a rate much faster than those of local cluster (near by galaxies). This unusual expansion was attributed to some form of *dark energy* (Turner, 1997) coming through the cosmological constant Λ. The idea was that as type Ia supernovae have almost the same intrinsic brightness (a standard candle), and since objects that are further away appear dimmer, the observed brightness of these supernovae can be used to measure the distance to them. The distance can then be compared to the supernovae's cosmological redshift which measures how much the universe has expanded since the supernova occurred; the Hubble law established that the further away an object is, the faster it is receding. The unexpected result was that objects in the universe are moving away from one another at an accelerating rate. Cosmologists at the time expected that recession velocity would always be decelerating, due to the gravitational attraction of the matter in the universe. (wikide). For a detailed review on this aspect one could refer to Peebles and Ratra (2002).

It is indeed fascinating and ironical to see that the cosmological constant Λ, that Einstein adopted to keep the universe static in 1917 but then discarded after the discovery by Hubble about the expanding universe, *has to be brought back in 1997 to keep the Universe accelerating.* As the continuous expansion requires energy, where does this energy come from? It is interpreted as the vacuum energy coming from the zero point energy as interpreted in the quantum field theory. Just as the critical density of baryonic matter is referred to as Ω_m the critical densities for the cdms and the cosmological constant are denoted by Ω_{cdm}, and Ω_Λ (FM 06). The constrained values of the above mentioned parameters as summarised by Lineweaver (1998) through the variety of sources like CMBR, distant supernovae and the mass-luminosity of clusters of galaxies are as follows:

$$\Omega_m = (\Omega_b + \Omega_{cdm}) = 0.24 \quad \text{and} \quad \Omega_\Lambda = 0.62.$$

These constraints were confirmed largely from the data obtained by WMAP too.

Acknowledging the fact that all the above discussions are based primarily on the most popular 'big bang theory', it is tempting to ask whether its rival the steady state theory has tried to explain any of these issues? In 2007, Narlikar *et al.* (2007) have put forward arguments to discuss the properties of the quasi-steady state cosmological model (QSSC) developed in 1993 in its role as a cyclic model of the universe driven by a negative energy scalar field. It was also discussed that the origin of such a scalar field in the primary creation process was first described by Hoyle & Narlikar in the early sixties. It is shown that the creation processes which take place in the nuclei of galaxies are closely linked to the high energy and explosive phenomena, which are commonly observed in galaxies at all redshifts. As they further point out the currently believed acceleration of the universe was already inherent in the earlier mentioned quasi steady state model as also the role of a scalar field. As the authors mention, *'The introduction of dark energy is typical of the way the standard cosmology has developed; viz., a new assumption is introduced specifically to sustain the model against some new observation. Thus, when the amount of dark matter proved to be too high to sustain the primordial origin of deuterium, the assumption was introduced that most of the dark matter has to be non-baryonic. Further assumptions about this dark matter became necessary, e.g., cold, hot, warm, to sustain the structure formation scenarios. The assumption of inflation was introduced to get rid of the horizon and flatness problems and to do away with an embarrassingly high density of relic magnetic monopoles. As far as the dark energy is concerned, until 1998 the general attitude towards the cosmological constant was typically as summarized by Longair in the Beijing cosmology symposium: 'None of the observations to date require the cosmological constant' (Longair 1987). Yet, when the supernovae observations could not be fitted without this constant, it came back with a vengeance as dark energy'*

As the observational technology improves it appears that the universe will provide more and more issues that have to be included and models reconstructed. Will there ever be an end to this or the play between the observations and theories continue from generation to generation, and may become a never answered question.

Amidst all these vagaries of Cosmic truth one is tempted to ask could 'dark matter' be a completely formless entity which may not follow any of the currently understood physics of normal matter and thus trying to discover it through the established physics may not be the way as all attempts made so far seem to bring in only upper or lower limits for its decay products, while no observational evidence is still identified as that of dark matter.

References

Abbott, B.P. *et al.* (2016). LIGO & VIRGO collaboration. *Phys. Rev. Letts.*, **116**, 061102.

Abramovici, A.A., *et al.* (I1992). *Sci.*, **256**, 325.

Abramowicz, M.A., Calvani, M. and Nobili, L. (1980). *Astrophys. J.*, **242**, 772.

Abramowicz, M.A., Carter, B., and Lasota, P. (1988). *GRG Journal*, **20**, 1173.

Abramowicz, M.A., Czerny, B., Lasota, J.P., and Szuszkiewicz, E. (1988). *Astrophys. J.*, **332**, 646.

Abramowicz, M.A., Jaroszynski, M., Sikora, M. (1978). *Astron. Astrophys.*, **63**, 221.

Abramowicz, M.A. and Marsi, C. (1987). *Observatory*, **107**, 245.

Abramowicz, M.A. and Miller, J.C. (1990). *MNRAS*, **245**, 729.

Abramowicz, M.A. and Prasanna, A.R. (1990). *MNRAS*, **245**, 720.

Abramowicz, M.A. Nurowski, P., and Wex, N. (1993). *CQ.G*, **10**, L183.

Alpar *et al.* (1982) *Nature*, **300**, 728.

Alpar M A *et al.* (1998) "Proceedings of the many faces of the Neutron Star" Kulwer.

Alpher, R.A. and Herman, R.C. (1948). *Nature*, **162**, 774.

Alpher, R.A. Bethe, H. Gamow, G. (1948). *Phys. Rev.*, **73**, 803.

Anderson, J.L. (1972). *Principles of Relativity Physics*, Academic Press.

Anderson, M.R. and Lemos, P. (1988). *MNRAS*, **233**, 489.

Baade, W. and Minkowski, R. (1954). *Astrophys. J.*, **119**, 215–231.

Baade, W. and Zwicky, F. (1934). *Phys. Rev.* **45**, 138.

Bachus *et al.* (1882) Ap J lets, 255, L63.

Banerjee *et al.* (1995), Ap.J., **449**, 789.

243

Banerjee *et al.* (1997) Ap.J, **474**, 389.

Baker Jr., R.M.L. (1967). *Astrodynamics: Applications and Advanced Topics.* Academic Press, New York, NY.

Bardeen, J. (1972). 'BlackHoles-les Astra Occlus' Eds Bryce and Cecil Dewitts.

Balbus, S. and Hawley, J.F. (1991). *Ap. J.*, **376**, 214.

Bardeen, J. (1970). *Astrophys. J.*, **162**, 71.

Bardeen, J., Press, W.H., and Teukolsky, S.A. (1972). *Astrophys. J.*, **178**, 347.

Begelman, M.C., Blandford, R.D., and Rees, M.J. (1984). *Rev. Modern Phys.*, **56**(2) Part 1.

Begelmann, *et al.* (1984). *Rev. Mod. Phys.*, **56**(2) Part 1.

Bethe, H.A. (1939). Energy Production in Stars. *Phys. Rev.*, **55**(5), 434–456.

Bhaskaran, P., Tripathy, S.C., and Prasanna, A.R. (1990). *J. Astrophys. Astr.*, (11), 461.

Bhaskaran, P. and Prasanna, A.R. (1990). *J. Astrophys. Astr.*, **11**, 49.

Bhaskaran P. and Prasanna, A.R. (1989). *Astrophys & Sp.Sci.*, **159**, 109.

Bisnovuaty Kogan, G.S. and Blinnikov, S.I. (1972). *Astrophys. Sp. Sci.*, **19**, 110.

Blanchet, L. (2014). *Living Reviews in Relativity*, **17**(2).

Blandford R and Narayan R (1986) Ap.J, **310**, 568.

Blanford, R.D., Netzer, H., and Woltjer, L. (1990). *Active Galctic Nuclei.* SaasFee Advanced Course, 20 Springer, Berlin.

Bolton, J., Stanley, G., and Slee, B. (1949). *Nature*, **164**, 101.

Bolton, J.G., Stanley, G.J., and Slee, O.B. (1949). *Nature*, **164**, 101–102.

Bondi, H. (1952). *MNRAS*, **112**, 195.

Bondi, H. (1952). *MNRAS*, **112**, 195.

Boyer, R.H. and Lindquist, R.W. (1967). *J. Math. Phys.*, **8**, 265.

Burbidge. E.M. Burbidge G.R., Fowler, W.A., and Hoyle F. (1957). *Rev. Mod. Phys*, **29**, 547. (also referred as B^{2}FH, 57)

Camenzind, M. (2008). *Compact Objects in Astophysics.* p. 513.

Camenzind Max (2007). *Compact Objects in Astrophysics*, Springer Nature.

Centrella, J., Baker, J., Kelly, B.J., and van Meter, J.R. (2010). *Rev. Mod. Phys.*, **82**, 3069, https://doi.org/10.1103/RevModPhys.82.3069

Chandrasekhar, S. (1935). *MNRAS*, **95**, 207.

Chandrasekhar, S. (1935). *MNRAS*, **95**, 207–255.

Chandrasekhar, S. (1939). *An Introduction to the Study of Stellar Structure.* University of Chicago Press, Illinois.

Chandrasekhar, S. (1964). *Astrophys. J.*, **140**, 417.

Chandrasekhar, S. (1964). *Phys. Rev. Letts.*, **12**, 114.

Chandrasekhar, S. (1964). *Phys. Rev. Letts.*, **12**, 437, *Astrophys. J.*, **140**, 417–433.

Chandrasekhar, S. (1992). *The Mathematical Theory of Blackholes*. Clarendon Press Oxford.

Cutler, C. Finn, L.S. Poisson, E., and Sussman, G.J. (1993). *Phys. Rev.*, **D47**, 151, *Phys. Rev. Letts.*, **70**, 2984.

Damour, T., Iyer, B.R., and Satyaprakas, B.S. (1998). *Phys. Rev.*, **D57**, 885.

Damour, T. and Taylor, J.H. (1991). *Ap, J.*, **366**, 501.

deSitter, W. (1917). *Proc. Roy. Akad. Sci. Amsterdam*, **19**, 1217.

Dicke, R.H., Peebles, P.J.E., Roll, P.G. and Wilkinson, D.T. (1965). Cosmic black-body radiation. *Ap. J.*, **142**, 414–419, https://doi.org/10.1086/148306.

Eddington, A. (1920). The internal constitution of the stars. *Nature*, **106**, 14–20.

Eddington, A.S. (1922). The theory of Relativity and its influence on scientific thought. Paperback Edition/, Amazon.

Eddington, A.S. (1924). *Nature*, **133**, 192.

Ehlers, J., Prasanna, A.R., and Breuer, R.A. (1987). *Class. Quantum. Grav.*, **4**, 253.

Ehlers, J. and Prasanna, A.R. (1996). *Class. Quantum. Grav.*, **13**, 2231.

Einstein, A. (1905). Zur Electrodynamic bewegte Korper. *Annalen der Physik*, 17.

Einstein, A. (1917). *Siz. Preuss. Akad. Wisse.* **142**, also *The Principle of Relativity* (Dover Publ. 1923, p. 35).

Eisberg, J.M. and Taylor, J.H. (2005). *The Relativistic Binary Pulsar B 1913+16 Thirty Years of Observation and Analysis*. Aspen Center Proceedings Rasio & Stairs (eds.). Vol. **328**, p. 2532.

en.wikipedia.org/wiki/Weakly_interacting_massive_particle#:~:text=There%20 exists%20no%20formal%20definition,also%20nonvanishing%20in%20 strength.

Eotvos, R.V. (1889). *Math. Naturwissen. Berlin aus Ungarn*, **8**, 63.

Esposito, L.W. and Harrison, E.R., (1975). *Ap. J. Letts.*, **196**, L.

Finkelstein, N. (1958). *Phys. Rev.*, **110**, 965.

Finkelstein, N. (1958). *Phys. Rev.*, **110**, 965.

Fishbone, L.G. and Moncrief, V. *Astrophys. J.*, **207**, 962.

Fowler, R.H. (1926). *MNRAS*, **87**, 114.

Frank, J., King, A., and Raina, D. (1985). *Accretion power in Astrophysics* Cambridge University Press.

Frank, J., King, A.R. and Raine, D. (2002). *Accretion Power in Astrophysics*, Cambridge University Press.

Freedman, W. (2021). *Ap. J.*, **919**(1), 16.

Friedmann, A. (1922). *Z. Phys.*, **10**, 377.

Friedmann, A. (1924). *Z. Phys.*, **21**, 326.

Galeev, A.A. Rosner, J., and Vaina, G., *Astrophys. J.*, **229**, 318.

Gamow, G. (1966). *Thirty Years that Shook Physics*. Dover Publications, New York, NY.

Ghosh, P. and Lamb, F.K. (1979). *Astrophys. J.,* **234**, 296.

Ginzburg, V.L. and Zeleznikov, V.V. (1975). *Ann. Revi. Astronomy Astrophys.* **13**, (A76-10076 01-88), *Palo Alto, Calif., Annual Reviews, Inc.*, 511–535.

Glendenning N K (2000), "*Compact Stars, Nuclear Physics, Particle Physics and General Reativity*", Springer.

Gold, T. (1978). *Nature*, **218**, 731.

Goldreich, P. and Julian. (1969). Pulsar Electrodynamics, *Ap. J.*, **157**, 869.

Goldstein, S. (1960). *Lectures on Fluidmechanics*. Interscience Publishers, London.

Gonthier and Harding. (1994). *Astrophysical J.*, **425**, 767.

Gorbukov, D.S. and Rubakov, V.A. (2011). *Introduction to the Theory of Early Universe*. World Scientific Publishing Company, Singapore.

Goswami, G., Chakravarty, G.K., Mohanty, S., and Prasanna, A.R. (2017). *Phys. Rev.*, **D95**, 103509.

Greenstein, J.L. and Munch, G. (1961). *Ann. Rpt. Dir. Mt.Wilson and Palomar Obs.* 80.

Greenstein, J.L. and Schmidt, M. (1964). *Astrophy. J.*, **140**, 1–34.

Hahn, S.G. and Lindquist, R.W. (1964). *Ann. Phys.*, **29**, 304.

Harrison, B.K., Thorne, K.S., Wakano, and Wheeler, J.A. (1965). *Gravitation Theory and Gravitational Collapse*. University Chicago Press.

Hartle, J. and Thorne, K.S. (1969). *Astrophys, J.*, **158**, 719.

Hartle, J.B. and Sharp, D. (1967). *Astrophys. J.*, **147**, 317.

Harwitt, M. (1973). *Astrophysical Concepts*. John Wiley.

Harwitt, M. (1982). Astrophysical concepts.

Hayashi, C. and Nakano, T. (1965). *Progr. Theoret. Phys. (Kyoto)*, **33**, 554; Hayashi, C. and Nakano, T. (1965). *Progr. Theoret. Phys. (Kyoto)*, **34**, 754; Hayashi, C. (1966). *Ann. Rev. Astro. Astrophys.*, **4**, 171–192.

Hazard, C., Mackey, M.B., Shimmins, A.J. (1963). *Nature*, **197**, 1037.

Hewish, A. (1951). https://en.wikipedia.org/wiki/Interplanetary_scintillation.

Hewish, A. (1951). *Pro. Roy. Soc.*, **A209**, 1096.

Hewish, A., Bell, S.J., Pilkington, J.D., Scott, P.F. and Collins, R.A. (1968). Observation of a rapidly pulsating radio source. *Nature*, **217**, 709–713.

Hewitt, J.N. *et al.* (1988). *Nature*, **333**, 537.

Hoyle, F., Fowler, W.A., Burbidge, E.M., and Burbidge, G.R. (1964). *Ap. J.*, **139**, 909.

Hoyle, F. and Fowler, W.A. (1963). *MNRAS*, **125**, 169, *Nature*, **197**, 533.

Hoyle, F. and Lyttleton, R.A. (1939). *Proc. Camb. Phil. Soc.*, **35**, 405.

https://en.wikipedia.org/wiki/Accelerating_expansion_of_the_universe.

Hulse, R.A. and Taylor, J.H. (1975). *Astrophys. J.*, **195**, L51.

Ichimaru, S. (1977). *Astrophys. J.*, **214**, 840.

Irvine, J.M. (1978). *Neutron Stars*. Clarendon Press, Oxford.

Jansky, K.G. (1933). Radio waves from outside the solar system. *Nature*, **132**(3323), 66.

Jeffrys, W.H. (1965). *Quasi Stellar Sources and Gravitational Collapse*, Robinson *et al.* (eds). p. 219.

Jones, T.J. *et al.* (2005). *Ap. J.,* **620**, 731, DOI:10.1086/427159.

Kameswar, W. (1991). *Chandra*, uni.Chicago press.

Kellermann Talk on Reber, ASP, July 1999. (1999). *NRAO/AUI Archives*, https://www.nrao.edu/archives/items/show/42029, accessed October 25, 2024.

Kerr, R.P. (1963). *Phys. Rev. Letts.*, **11**, 237.

Kolb, E.W. and Turner, M.S. (1989). *The Early Universe*, (Ind edn.). Levant Books, Kolkota.

Kolb, E.W. and Turner, M.S. (1989). *The Early Universe*. (Levant Books Kolkota (Indian Edition).

Kozlowski, M., Jaroszynski, M., and Abramowicz, M.A. (1978). *Astron. Astrophys.*, **63**, 209.

Landau, L.D. (1932). *Physik. Zeits. Sovejtunion.* 1, **285**.

Landau, L.D. and Lifshitz, E.M. (1951). *Classical Theory of Fields*. Pergamon Press.

Lang, K.R. (1974). *Astrophysical Formulae*. Springer-Verlag, New York, NY.

Lemaitre, G. (1927). *Ann. Soc. Sci. Brux.*, **A47**, 49.

Lemaitre, G. (1931). *MNRAS* **91**, 483.

Lightman, A.P. (1974). *Ap. J.*, **194**, 420.

Lightman, A.P., Shapiro, S., and Rees, M.J. (1978). *Physics and Astrophysics of Black holes and Neutron Stars*. North Holland.

Lineweaver, C.H. (1998). The cosmic background and observational convergence in the WM-WL plane. *Ap. J.*, **505**, L69.

Liu *et al.* Detailed study of a rare hyperluminous rotating disk in an Einstein Ring 10 billion years ago, https://doi.org/10.1038/s41550-024-02296-7.

Longair, M.S. (1987). *IAU Symposium 124, "Observational Cosmology.* (eds.), Hewitt, A., Burbidge, G., Fang, and L. Z., Reidel, D., (eds.), Dordrecht, p. 823.

Lord Rayleigh, R. (1917). *Proc. Roy. Soc.*, **A93**, 148.

Lorentz, H.A. (1892). *Neerl*, **25**, 363; Versl, K., Amsterdam, (1904). **1**; 74; *Proc. K. Ak. Amsterdam* **6**, 809.

Luyten, W.J. (1963). *Pub. Astr. Obs. Minnisota*, **111**(13).

Lynden Bell, D. (1969). *Nature*, **226**, 64.

Lyndon Bell, D. (1969). *Nature*, **226**, 64.

Maraschi, L., Reina, C., and Treves, A. (1987). *Asytrophys. J.*, **35**, 389.

Mashoon B *et al.* (1984). *GRG journal*, **16**(8), 711.

Mashoon, B., Hehl, F.W., and Theiss, D.S. (1984). *GRG J.*, **16**(8), 711.

Massey *et al.* (2015). *Mon. Not. R. Astron. Soc.*, **449**, 3393.

Mathews, P.M. and Venkatesan, K. (1976). *A Text Book of Quantum Mechanics*. Tata Mcgraw Hill Publishing Company, New Delhi.

Mathis, J.S. (2019). Planetary nebula. *Encyclopedia Britannica*, 26 February, https://www.britannica.com/science/planetary-nebula.

Matthews, T.A. and Sandage, A.R. (1963). *Ap. J.*, **138**, 30.

Maxwell, J.C. (1865). *Phil. Trans. Roy. Soc. (London)*, **155**, 450; *Sci. Papers*, **1**, 526–597.

May, M.M. and White, R.H. (1966). *Phys. Rev.*, **141**, 140.

McCrea, W.H. and Milne, E.A. (1934). Newtonian Universe and the curvature of space. *Quart. J. Mathematics*, **os-5** (1), 73–80.

Milne Thomson, L.M. ( ). *Theoretical Hydrodynamics*. Dover Publications.

Minkowski, H. (1908). Space and time. Address at the 80th Assembly of German Natural Scientists Meet at Cologne (September) reproduced in *The Principle of Relativity*. Dover Publications. 1952.

Misner, C.W. and Sharp, D.H. (1964). *Phys. Rev.*, **136 B**, 571.

Misner, C.W. Thorne, K.S., and Wheeler, J.A. (1973). *Gravitation*. Freeman and Co.

Mossbauer, R.L. (1958). *Z. Physik.*, **151**, 124; *Naturwissenshaften*, **45**, 538.

Mukhopadhyay, B. and Prasanna, A.R. (2003). *Int. J. Mod. Phys.*, **12**(1), 157, **18**(7), 1091.

Narayan, R. and Yi, I. (1994). **428**, L13 (1995), *Ap. J.*, **444**, 231, **452**, 710.

Narlikar, J.V., Burbidge, G., and Vishwakarma, R.G. (2007). Cosmology and cosmogony in a cyclic universe. *J. Astrophys. Astronomy*, **28**, 67–99.

Narlikar J.V. (1977). *The Structure of the Universe*. Oxford University Press.

Ohanian, H.C. and Ruffini, R. (1976). *Gravitation and Spacetime*. Norton & Company, New York, NY.

Ohanian and Ruffini, R. (1976). Gravitation & Spacetime.

Oppenheimer, J.R. and Snyder, (1939). *Phys. Rev.*, **56**, 455.

Oppenheimer, J.R. and Volkoff, G.M. (1939). *Phys. Rev.*, **55**, 374.

Ostriker, J. (1999). *Astrophys. J.*, Centennial Issue, **525C**, 297.

Pacini, F. (1968). *Nature*, **219**, 145–146, https://doi.org/10.1038/219145a0.

Paczynski, B. (1986). *Astrophys. J.*, **301**, 503.

Parker, E.N. (1958). *Astrophys. J.*, **128**, 664.

Peebles, J. (1965). *Ap. J*, **142**, 1317.

Peebles, J. and Ratra, B. (2002). The cosmological constant and dark energy, arxive: astro-ph/0207347v2.

Penrose, R. (1965). *Phys. Rev. Letts.*, **14**, 57.

Penrose, R. (1969). *Nuovo Cimento 1* (Special Number) 252.

Penrose, R. (1969). *Riv. Nuovo Cim.* **1**, 252–276, *GRG J.* (2002). **34**, 1141–1165.

Penzias, A.A. and Wilson, R.W. (1965). *Ap. J.*, **142**, 419.

Perlick, V. (2004). *Living Reviews in Relativity.* **7**(9).

Peters, P.C. and Mathews, J. (1963). *Phys. Rev.*, **131**, 435.

Peterson, J.A. (1975). *Phys. Rev. D.*, **12**, 2218.

Peterson, J.A. (1975). *Phys. Rev.*, *D.*, **12**, 2218; (1874), **D10**, 3166.

Pfister, H. (1995). *Mach's Principle-From Newton's Bucket to Quantum Gravity.* J. Barbour and H. Pfister (eds.), Birkhauser.

Piran, T. (1978). *Ap. J.*, **221**, 652.

Pirani, F.A.E. (1957). *Phys. Rev.*, **105**, 1089.

Planck, M. (1990). *Ann. Phys.*, **1**, 69; *Phys. Abh.*, Bd. I, S. 61.

Poincare, H. (1905). *C.R. Acad. Sci. Paris.*, **140**, 1504; (1906). *Rend. Circ. Mat. Palermo*, **21**, 129.

Pound, R. and Rebka, G.A. (1960). *Phys. Rev. Letts.*, **4**, 537.

Prasanna, A.R. (1973). *Int. J. Theo. Phys.*, **8**(4), 237.

Prasanna, A.R. (1980). Review Article *"Revista del Nuovo Cimento"*, **3**(11), 1–50.

Prasanna A.R (1984), in Proceedings of the workshop on "Gravitation and Relativistic astrophysics, (1982), Eds Prasanna, Narlikar and Vishveswara, World Scientific.

Prasanna, A.R. (1989). *Astron. Astrophys.*, **217**, 329.

Prasanna, A.R. (1999). *Phys. Letts.*, **A257**, 120.

Prasanna, A.R. (2002). *Mod. Phys. Letts.*, **A17**, 1835.

Prasanna, A.R. (2015). *J. Electromag. Waves App.*, **29**(3), 283–330.

Prasanna, A.R. (2017). *Gravitation.* CRC Press, Taylor & Francis Group.

Prasanna, A.R. and Bhaskaran, P. (1989). *Astrophys & Sp. Sci.*, **153**, 201.

Prasanna, A.R. and Chakrborty, D.K. (1980). *Pramana*, **14**, p. 113.

Prasanna, A.R. and Gupta, A. (1997). *Il Nuovo Cimento.*, **B112**, 1089.

Prasanna, A.R. and Sengupta, S. (1994). *Phys. Letts*, **A 193**, 25.

Prasanna, A.R. and Varma, R.K. (1977). *Pramana, J. Phys.*, **8**(3), 229.

Prasanna, A.R. and Vishveswara, C.V. (1978). *Pramana, J. Phys.*, **11**, 358.

Prasanna, R. and Dadhich, N. (1982). *Il Nuovo Cimento.*, **72**(1), 42.

Prenderghast, H. and Burbidge, G. (1968). *Astrophys. J.* (Letters), **151**, L 83.

Prenerghast, H. and Burbidge, G. (1968). *Astrophys. J.* (*Letters*), **151**, L83.

Pretorius, F. (2005). *Phys. Rev. Letts.*, **95**, 121101.

Pretorius, F. (2006). *Class. Quant. Grav.*, **23**, S. 5216, S. 529.

Pringle, J. and Rees, M. (1972). *Astr. Ap.*, **21**, 1.

Pringle, J. and Rees, M.J. (1972). *Astr. Ap.*, **21**,1.

Pringle, J. E. (1981). *Ann. Rev. Astr. Astrophys.*, **19**, 137.

Ray. (1992). *Introducing Einstein's Relativity.* Clarendon Press, Oxford.

Radhakrishnan V and Srinivasan G (1982) *Current Science,* **51**, 1096.

Rai Choudhuri, A. (1998). *The Physics of Fluids and Plasma.* Cambridge University Press.

Rees, M. (1997). *Before the Beginning.* Perseus Books, Cambridge.

Robertson, H.P. (1935). *Ap. J.*, **82**, 284 (1936). **83**, 187, 257.

Roll, P.G., Krotokov, R., and Dicke, R.H. (1964). *Ann. Phys. USA,* **26**, 442.

Rubin, V. (1988). *Proc. Am. Phil. Soc.*, **132**(4).

Ryden, B. (2017). Introduction to Cosmology.

Ryle, M., Smith, F.G. and Elsmore, B. (1950). *MNRAS*, **110**(6), 508–523, https://doi.org/10.1093/mnras/110.6.508.

Sakharov, A.D. (1967). *JETP Letts.*, **5**, 24–27.

Salpeter, E.E. (1964). *Astrophys J.*, **140**, 796.

Sandage, A. (1965). *Quasi Stellar Sources and Gravitational Collapse.* Robinson *et al.* (eds.), p. 265.

Satyaprakash, B.S. and Schutz, B.F. (2009). *Living Reviews in Relativity*, **12**, 2009.

Schmidt, G. (1966). *Physics of High Temperature Plasma.* Academic Press Inc. New York, NY.

Schmidt, M. (1963). *Nature*, **197**, 1040.

Schnider, P., Ehlsrs, J., and Falco, E.E. *Gravitational Lenses*, Springer, New York, NY.

Seguin, F.H. (1975). *Astrophys. J.*, **197**, 745.

Shakura, N.I. and Sunyaev, R.I. (1973). *Astr. Ap.*, **23**, 337.

Shapiro, S.L. and Teukolsky, S.A. (1983). *Blackholes, White Dwarfs and Neutrons Stars*, John Wiley.

Shibazaki, and Hoshi, R. (1976). *Prog. Theo. Phys.*, **54**, 706.

Shields, G.A. (1999). *The Publications of the Astronomical Society of the Pacific, Volume* 111, Issue 760, pp. 661–678.

Shvartzman, V.F. (1971). *Soviet Astr.*, **15**, 37.

Smith, F.G. (1950). *Nature*, **165**, 422–423.

Sunyaev, R.A. and Zeldoich, Y.A.B. (1970). *Astrophys. Space Sci.* **7**, 3; (1972). *Comm. Astrophys. Space Phys.*, **4**, 173.

Thorne, K.S. (1971). *General Relativity & Cosmology*, 47th International School Fermi, Sachs R.K.)(ed.), pp. 238–283.

Thorne, K.S. (1987). 300 Years of Gravitation. Hawking & Israel (eds.). Cambridge University Press.

Treves, A., Maraschi, L., and Abramowicz, M.A. (eds.). (1989). *Accretion*. World Scientific.

Tripathy, S.C., Dwivedi, C.B., Das, A.C., and Prasanna, A.R. (1993). *J. Astrophys. Astr.*, **14**, 103 and 167.

Tripathy, S.C., Prasanna, A.R., and Das, A.C. (1990). *MNRAS*, **240**, 384.

Tripathy, S.C., Prasanna, A.R., and Das, A.C. (1990). *MNRAS*, **246**, 384.

Turner, M.S. and White, M. (1997). *Phys. Rev.*, **D56**, R4439.

Turner, M.S. and Widrow, L. (1988). *Phys. Rev.*, **D37**, 2743.

Vekatraman, G. (1992). Chandrasekhar limit.

Vishveswara, C.V. (1970). *Nature*, **227**, 936; *Phys. Rev.*, **D1**, 2870.

Wagoner, R. (1975). *Ap. J. Letts.*, **196**, L63.

Walker, A.G. (1936). *Proc. Lond. Math. Soc.*, (2), **42**, 90.

Walsh, D. Carswell, R.F. and Wymann, R.J. (1979). *Nature*, **279**, 381.

Weber, J. (1966).

Weinberg, S. (1972). Gravitation and Cosmology.

Weinberg, S. (2007). Cosmology.

Wiita, P. (1985). *Comm. Astrophys.*, **10**(5), 199.

Woltjer, L. (1964). *Astrophys. J.*, **140**, 1309.

York, J.W.J. (1983). *Gravitational Radiation*, p. 175. N. Deruelle and Piran, T. North Holland, Amsterdam.

Zeldovich, Y.A.B. (1964). *Soviet. Phys. Doklady*, **9**, 195.

Zeldovich, Y.A.B. and Novikov, I.D. (1971). *Relativistic Astrophsics, I Stars and Relativity*. University Chicago Press.

Knobloch, E. (1992). *MNRAS*, **255**, 25.